T0205491

Integrated Hybrid Resonant DCDC Converters

Peter Renz • Bernhard Wicht

Integrated Hybrid Resonant DCDC Converters

 Springer

Peter Renz
ASIC Design Unit
Robert Bosch GmbH
Reutlingen, Germany

Bernhard Wicht
Institute of Microelectronic Systems
Leibniz University Hannover
Hannover, Germany

ISBN 978-3-030-63946-4   ISBN 978-3-030-63944-0   (eBook)
https://doi.org/10.1007/978-3-030-63944-0

This Springer imprint is published by the registered company Springer Nature Switzerland AG
The registered company address is: Gewerbestrasse 11, 6330 Cham, Switzerland

Preface

This book explores the solutions on system and circuit level for hybrid resonant converters for portable applications operating from a Li-Ion battery (3.0–4.5 V) at power levels of a few hundred milliwatts. In particular, the growing field of wearables, for example, for medical monitoring devices and smartwatches, requires highly compact power supplies with high energy efficiency. The final goal for these space-constrained applications are fully integrated DCDC converters, which include all passive components either on-chip or by co-integration in the same package. Highly integrated state-of-the-art system-in-package solutions comprise inductive step-down converters, which operate in the lower MHz range with passive components in the order of μF and μH. In order to further scale down the size of passives, new conversion concepts are required. Hybrid switched-capacitor (SC) DCDC converters pursue a new and extremely promising approach by combining capacitor-based and inductive concepts in a single converter structure. Resonant operation of these hybrid converters allows for switching frequencies in the multi-Megahertz range (>10 MHz) at significantly reduced dynamic losses.

The objective of this book is to provide a systematic and comprehensive insight into design techniques for hybrid resonant switched-capacitor converters. Written in handbook style, the book covers the full range from fundamentals to implementation details including topics like power stage design, gate drive schemes, different control mechanism for resonant operation, and integrated passives. The material will be interesting for design engineers in industry as well as researchers who want to learn about and implement advanced power management solutions.

The main topics are as follows: (1) A new multi-ratio resonant converter architecture is introduced, which enables lower switching frequencies and better passive component utilization. This leads to high power efficiency as well as to full integration of all passive components. (2) The circuit block design for high efficiency of the power stage is investigated. (3) Implementation details and concepts for integrated passives are explored. (4) Different control mechanisms are derived, modeled, implemented, and compared to each other.

In particular, this book presents the first fully integrated resonant SC converter, which is extended by a multi-ratio power stage in order to cover the wide Li-Ion input voltage range of 3.0–4.5 V. The design achieves a peak efficiency of 85% with an integrated 10 nH on-chip planar inductor (fully integrated) and 88.5% with a 10 nH in-package inductor (highly integrated).

Different control mechanisms are investigated and implemented by a fast mixed-signal controller. Switch conductance regulation (SwCR) operates at resonance frequencies as large as 47 MHz, which offers full integration of passives on IC level (no external components). In contrast, resonant bursting with dynamic off-time modulation (DOTM) achieves higher efficiency at the cost of an external output capacitor. The light load case down to currents as low as 0.5 mA is supported by automatic transition into pure non-resonant SC mode, controlled by frequency modulation.

Since there is no dynamic model available, yet, which allows to analyze the stability for this class of resonant SC converters, a model for the nonlinear mixed-signal control loop of the switch conductance regulation (SwCR) is proposed. The model enables stability analysis and allows for an optimal design of the control loop.

For the implementation of the power stage, the efficiency benefit of low-voltage transistor stacking over single high-voltage switches is investigated with a detailed but easy-to-use model of the transistor stack. A new implementation option for stacking of three low-voltage transistors for higher voltage capability is presented, which is independent of the input voltage. Finally, different concepts, implementation details, and achievable parameters of integrated passives (capacitors and inductors) for resonant SC converters are presented.

This book is based on our research at the Institute for Microelectronic Systems at Leibniz University Hannover, Hannover, Germany, in cooperation with Texas Instruments, Freising, Germany. We are grateful to many team members at the university as well as at our industry partner. Special thanks goes to Michael Lueders, Texas Instruments, for his constant interest in the topic and continuous support. He was always available for in-depth technical discussions, and his expertise in analog and power design was precious and helpful. We would like to convey many thanks to Dominique Poissonnier, Giovanni Frattini, and Erich Bayer for their invaluable advice and comments. We also want to thank Sebastian Beringer and Marc Christopher Wurz, Institute of Micro Production Technology at Leibniz University Hannover, for providing different microfabricated inductors for comparison and evaluation. We also appreciate the contributions of many students, especially the excellent work of Maik Kaufmann and Niklas Deneke.

A special thanks goes to our families, without their love and support this book would not have been possible.

Gomaringen, Germany   Peter Renz
Gehrden, Germany   Bernhard Wicht
September 2020

Contents

Acronyms

List of Abbreviations

BCD	Bipolar-CMOS-DMOS process technology
CMOS	Complementary metal-oxide-semiconductor
CPU	Central processing unit
DOTM	Dynamic off-time modulation
EEF	Efficiency enhancement factor
EMI	Electromagnetic interference
EME	Electromagnetic emission
ESR	Equivalent series resistance
FCML	Flying capacitor multilevel converter
FSL	Fast-switching limit
GMD	Geometric mean distance
GPU	Graphical processor unit
IoT	Internet of things
LDO	Low dropout regulator
MEMS	Micro-electrical-mechanical systems
MIM	Metal insulator metal
MOM	Metal-oxide-metal
MOS	Metal-oxide-semiconductor
NMOS	Negative-channel metal-oxide-semiconductor field-effect transistor
PCB	Printed circuit board
PMIC	Power management IC
PMOS	Positive-channel metal-oxide-semiconductor field-effect transistor
PMU	Power management unit
PWM	Pulse width modulation
QFN	Quad flat-no leads
ReSC	Resonant switched-capacitor converter
RMS	Root mean square
SC	Switched-capacitor converter

SiP	System-in-package	
SMD	Surface-mounted device	
SSL	Slow-switching limit	
SoC	System-on-chip	
SwCR	Swichted conductance regulation	
ZCS	Zero current switching	

List of Symbols

A_{Cfly}	m^2	Area consumption of flying capacitors
A_{eff}	m^2	Effective cross section area of bond wire
A_L	m^2	Area consumption of integrated inductor
α	F	Quality factor of integrated capacitor
C_{GS}	F	Gate-source capacitance
C_{GD}	F	Gate-drain capacitance
C_{DS}	F	Drain-source capacitance
C_i	F	Equivalent capacitance value in phase i
$C_\square$	F/m^2	Capacitance density
C_{ox}	F/m^2	Oxide capacitance per area
C_p	F	Pump capacitor
C_{eq}	F	Energy-based equivalent capacitance
C_{fly}	F	Flying capacitor
C_{fly1}	F	Flying capacitor 1
C_{fly2}	F	Flying capacitor 2
C_{fly3}	F	Flying capacitor 3
C_{BP}	F	Parasitic bottom plate capacitor
C_{SUB}	F	Parasitic substrate capacitance of inductor
C_{TP}	F	Parasitic top plate capacitor
C_{in}	F	Input capacitor
C_{out}	F	Output capacitor
C_{ox}	$\frac{F}{m}$	Oxide capacitance per area
δ	µm	Skin depth
$DLCR$		Dynamic load current range
E_i	J	Energy dissipation in phase i
ϵ_o	$\frac{F}{m}$	Permittivity of vacuum
$\epsilon_{r,oxide}$	$\frac{F}{m}$	Relative permittivity of silicon dioxide
f_{clk}	Hz	Clock frequency of the oscillator
f_{sw}	Hz	Switching frequency
$f_{sw,res}$	Hz	Resonance frequency
$f_{sw,res,1/2}$	Hz	Resonance frequency in ratio 1/2
$f_{sw,res,1/3}$	Hz	Resonance frequency in ratio 1/3
$f_{sw,res,2/3}$	Hz	Resonance frequency in ratio 2/3

$F_{\text{counter}}(s)$		Transfer function of the digital counter
F_{linears}		Combined transfer function of all linear components
$F_{\text{ReSC}}(s)$		Plant transfer function of ReSC converter
GM		Gain margin
GMD	m	Geometric mean distance
GND	V	Ground, global reference potential
G_{sw}	S	Switch conductance of the power switch
G_{LSB}	S	Least significant bit of the switch conductance
G_{offset}	S	Offset of the switch conductance
$G_{\text{sw,OP}}$	S	Conductance of the power switch at operating point
G_{LSB}	S	Smallest unity conductance value
I_i	A	Current in phase i
I_{in}	A	Input current of the converter
I_{RMS}	A	RMS current
I_{L}	A	Inductor current of the converter
$\hat{I}_{\text{L}}$	A	Inductor peak current of the converter
I_{out}	A	Output current of the converter
$I_{\text{out,max}}$	A	Maximum output current of the converter
$I_{\text{out,min}}$	A	Minimum output current of the converter
$I_{\text{out,OP}}$	A	Output current of the converter in the operating point
K_{Gsw}	S	Proportional term of the segmented power switches
K_{corr}		Correction factor for inductor
K_{Rout}	Ω	Proportional term for the equivalent output resistance
L	H	Inductor of resonant converter
L	m	Length of transistor
L_{s}	H	Self-inductance of a planar inductor
L_{model}	H	Inductance for dynamic model
l_{winding}	H	Total length of planar inductor
l_{w}	H	Side length of bond wire
M	H	Mutual inductance of wire segment
M_{+}	H	Positive mutual inductance
M_{-}	H	Negative mutual inductance
η		Efficiency
η_{CP}		Efficiency of charge pump
N		Conversion ratio
$N\left(\hat{x}_{\text{in}}\right)$		Describing function
P_{CBP}	V	Losses introduced by parasitic bottom plate capacitor
η_{LR}		Efficiency of linear regulator
η_{SC}		Efficiency of switched-capacitor converter
P_{cond}	W	Conduction losses of the power switches
P_i	W	Power dissipation in phase i
P_{loss}	W	Total power loss of the converter
P_{Rout}	W	Power loss due to equivalent output resistance
P_{sw}	W	Switching losses of the power switches
$P_{\text{sw,tot}}$	W	Total switching losses of the power switches

P_{tot}	W	Total losses of power switch
$P_{LS,tot}$	W	Total losses of the level shifters
P_{in}	W	Input power of the converter
P_{out}	W	Output power of the converter
PM		Phase margin
ρ	Ω m	Electrical resistivity
Q	V	Quality factor of inductor
r_{bond}	m	Radius bond wire
R_{BP}	Ω	Bond pad resistance
$R_{DS,on}$	Ω	On-resistance between drain and source
R_i	Ω	Equivalent resistance in phase i
R_{sw}	Ω	Switch resistance of the power switch
R_{offset}	Ω	Offset resistance of the power switch
$R_{sw,OP}$	Ω	Switch resistance of the power switch
R_{ESR}	Ω	Equivalent series resistance of inductor
R_{out}	Ω	Equivalent output resistance
$R_{out,SC}$	Ω	Equivalent output resistance in SC operation
$R_{out,SSL}$	Ω	Equivalent output resistance in slow-switching limit
$R_{out,FSL}$	Ω	Equivalent output resistance in fast-switching limit
$R_{out,DOTM}$	Ω	Equivalent output resistance in DOTM
$R_{out,OP}$	Ω	Equivalent output resistance in the operating point
T_{clk}	s	Period of the clock frequency of the oscillator
T_I	s	Integration time constant of the digital counter
$t_{winding}$	Ω	Thickness of the winding of the planar inductor
$t_{winding,eff}$	Ω	Effective thickness of winding
t_{ox}	m	Thickness of the oxide layer
V_{boot}	V	Bootstrapped voltage
VCR		Voltage conversion ratio
V_{blk}	V	Blocking voltage
V_{DD}	V	Positive supply voltage
V_{fb}	V	Feedback voltage for the control
V_{GS}	V	Gate-source voltage
V_{DS}	V	Drain-source voltage
V_i	V	Equivalent voltage across capacitor in phase i
V_{in}	V	Input voltage of the converter
V_{ref}	V	Reference voltage
$V_{SS,HS}$	V	High-side ground
$V_{DD,HS}$	V	Positive high-side supply voltage
$V_{DD,LS}$	V	Positive low-side supply voltage
V_{out}	V	Output voltage
V_{sw}	V	Voltage at the switching node
V_{th}	V	Threshold voltage of a transistor
W	m	Width of transistor
$w_{winding}$	m	Winding width of inductor
φ		Clock signal of the oscillator

φ_{CP}		Clock signal for the charge pump
φ_{SC}		Clock signal in SC mode
$\varphi 1_{SC}$		Clock phase one in SC mode
$\varphi 2_{SC}$		Clock phase two in SC mode
φ_{DOTM}		Clock signal in DOTM mode
$\varphi 1_{DOTM}$		Clock phase one in DOTM mode
$\varphi 2_{DOTM}$		Clock phase two in DOTM mode
$\varphi 1$		Clock phase one
$\varphi 2$		Clock phase two
φ_{CP}		Charge pump clock signal
μ_n	$\frac{cm^2}{Vs}$	Effective electron mobility
μ_o	$\frac{Vs}{Am}$	Magnetic permeability of free space
ζ		Damping factor of a resonant circuit

Chapter 1
Introduction

1.1 Motivation

Wearables and portable devices are an integral part of the Internet of things (IoT) with its vision that nearly every device and service is connected to the Internet as indicated in Fig. 1.1. All "*things*" connect, communicate, and coordinate with each other. This will transform the way we work and live. The worldwide market for wearables is expected to grow at an annual rate (CAGR) of 11.3% from 2019 to 2025 to reach $62.82 billion by 2025 [1].

While currently most of the wearables are used in simple wrist-mounted fitness trackers and smartwatches, future use cases may include smart glasses [2], smart headphones [3], smart shoes [4], smart clothes (e-textiles) [5], smart tattoos [6], or medical sensors and implants [7]. Wearables are expected to play a major role in areas such as communication, health monitoring, augmented reality, and authentication.

The rapid development of wearables and IoT devices toward miniaturized and power-efficient electronics can be considered to be similar to the evolution of the integrated circuit. Semiconductor technology scaling has enabled exponential growth of the transistor density, which has lead to significantly smaller and faster computing power over the last 60 years [8].

Higher integration brings several significant benefits. More functionality can be added together with even smaller form factors. This is especially important for portable applications where the available space is limited. Due to the shorter distance between functional blocks, the parasitics are reduced, which improves the system speed and performance.

Especially in mobile devices, multi-chip approaches have evolved to system-on-chip (SoC) or system-in-package (SiP) solutions that integrate the CPU, graphical processing unit GPU, memory, and analog and digital blocks all on one chip (SoC)

P. Renz, B. Wicht, *Integrated Hybrid Resonant DCDC Converters*,
https://doi.org/10.1007/978-3-030-63944-0_1

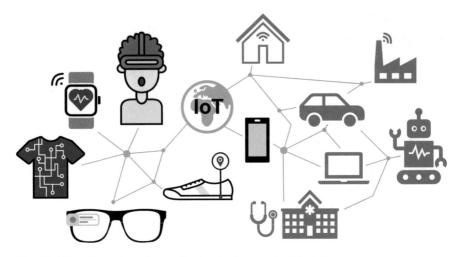

Fig. 1.1 Wearables and portable applications in the context of the IoT ecosystem

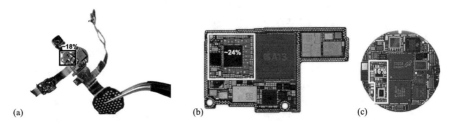

(a)   (b)   (c)

Fig. 1.2 Different main boards of wearable and portable devices. The occupied area by power management is marked with boxes. (**a**) Apple AirPods (smart headphone) [10]; (**b**) iPhone 11 Pro Max (smartphone) [11]; (**c**) Motorola Moto 360 (smartwatch) [12]

or in a single package (SiP). Thus, the main boards of the portable devices consist of fewer individual chips and passive components.

Power management units have resisted the integration and have shrunk in size only to a small extent [9]. They often consist of a power management IC (PMIC) with multiple integrated power stages, linear regulators (LDO), and control circuits. However, the passive components for energy storage, which are essential for the power conversion, are placed externally due to their large size. These passive components, mainly power inductors and capacitors, do not follow the scaling of Moore's Law [8] leading to a mismatch in future development.

Figure 1.2 shows different main boards of wearable and portable devices. While a lot of functionality (CPU, GPU, etc.) can be integrated in SoCs, the power management block (marked boxes in Fig. 1.2) still occupies a large PCB area, e.g., up to 24% in the iPhone 11 Pro Max. Especially the high number of external SMD inductors and capacitors around the PMIC is clearly visible in Fig. 1.2. Hence, the power management is still a major factor that determines the system size, which is a crucial parameter in wearable and portable applications.

Additional challenges arise for the power management systems since the number of voltage domains, the power demand, as well as the switching speeds will increase. This leads to a higher number of required DCDC converters. PMICs with external components require an increased pin count and degrade the performance due to interconnection parasitics and potential issues regarding electromagnetic interference (EMI).

The ultimate solution for these challenges is the full integration of the converters on a single chip. This can also reduce costs, increase reliability (due to lower component count and interconnections), and simplify the design for the user [13, 14]. The largest challenge for fully integrated conversion so far is the lack of high-quality inductors. Conventional converters all rely on at least one inductor. Only small inductance values are possible, which leads to very high switching frequencies, associated with high switching losses. This results in low efficiencies and limited voltage conversion ratios [15–21]. In recent years there is a growing interest in switched-capacitor (SC) converters due to their good integration capability in standard CMOS processes [22–32]. They can achieve high efficiencies at low output power, but they often do not support high-power operation. Moreover, SC converters do not have inherent regulation capability. Hybrid converters are a promising converter class that supports integration of inductive and capacitive components while minimizing losses and improving power density [33–41]. However, the requirements for the use in wearable and portable electronics are not met yet since low-power operation is often not supported and the input voltage range is limited (especially for fully integrated converters).

1.2 Scope and Outline of This Book

The scope of this work is summarized and depicted in Fig. 1.3. It is strongly related to the trend toward fully integrated power management with high efficiency, driven by portable and wearable IoT applications. DCDC converters are required for the down-conversion of a battery voltage to the supply voltages for both digital and analog circuits in processors and SoCs, which are continuously decreasing. This work covers fully integrated hybrid resonant DCDC converters in standard silicon technologies, which support a wide input voltage range of 3.0 to 4.5 V suitable for Li-Ion batteries and which also maintain high efficiency from high to low output power levels.

Requirements in terms of cost, small form factor, and better reliability are addressed by full integration of the power stage along with the required energy storage components such as capacitors and inductors. Very high switching frequencies (>100 MHz) are often used to scale down the size of the passive components, which in turn reduces the overall efficiency, especially at low output power levels.

In this work, the limitations of recent fully integrated DCDC converters are addressed by four main topics: (1) A new multi-ratio resonant converter architecture is introduced, which enables lower switching frequencies and better passive compo-

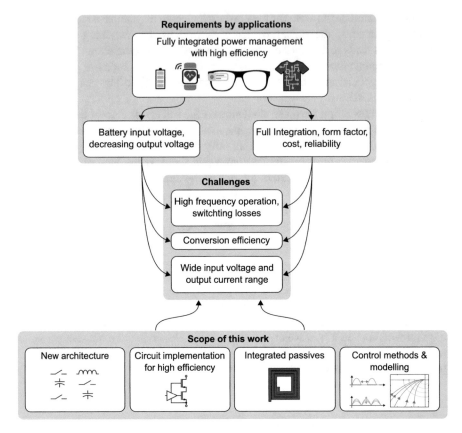

Fig. 1.3 Summary of the scope of this work

nent utilization. This leads to high power efficiency as well as to full integration of all passive components. (2) The circuit block design for high efficiency of the power stage is investigated. (3) Implementation details and concepts for integrated passives are investigated. (4) Different control mechanisms are implemented, modeled, and compared to each other.

The outline of this book is as follows. Chapter 2 highlights the motivation of this work and the demand for fully integrated power management for wearable and IoT applications. Section 2.2 gives an overview of different types of DCDC converters. Hybrid DCDC converters combine inductive and capacitive components while minimizing losses and improving power density. Different state-of-the-art DCDC converters with their pros and cons are compared. The hybrid resonant SC converter concept is the focus of this work since it shows the highest potential for fully integrated voltage conversion. The fundamentals of resonant SC converters are covered in Sect. 2.3.

The concept of hybrid multi-ratio resonant SC conversion is introduced in Chap. 3. Section 3.1 explains the basic topology for multi-ratio resonant conversion and the operation principle. Different control mechanisms depending on the size of the passive components are investigated in Sect. 3.2. An efficiency model for loss analysis and for finding of an optimal operation point is introduced in Sect. 3.3. Experimental results and a comparison to the state-of-the-art are presented in Sect. 3.4.

Chapter 4 explores the implementation and the control of the power switches, which is one of the key challenges for highly efficient converter operation. The efficiency benefit of low-voltage transistor stacking over single high-voltage switches is investigated in Sect. 4.1. Segmentation of the power switches, required for the proposed switch conductance regulation, is discussed in Sect. 4.2. Section 4.3 describes and evaluates different level shifter topologies. The generation of a flying gate drive supply is shown in Sect. 4.4, while Sect. 4.5 presents the full power stage implementation.

The implementation of the passive components plays an important role for fully integrated converters, covered in Chap. 5. Section 5.1 presents the design of loss optimized integrated flying capacitors. The modeling and implementation of integrated inductors are covered in Sect. 5.2. In addition, a comparison between fully integrated and off-chip inductors is shown.

Chapter 6 presents details on different control options for resonant SC converters. The implementation of two different control schemes is shown in Sect. 6.1 along with measured transient responses. A dynamic model of the resonant SC converter is introduced in Sect. 6.2, and the modeling of the nonlinear digital control loop for switch conductance regulation is discussed in Sect. 6.3. The model enables stability analysis and allows for an optimal design of the control loop. Experimental results verify the model.

Chapter 7 summarizes and concludes the results of this work.

References

1. PR Newswire: *World Market for Wearable Devices, Set to Reach $62.82 Billion by 2025 – Increasing Penetration of IoT & Related Devices Drives Market Growth* (2019). https://www.prnewswire.com/news-releases/world-market-for-wearable-devices-set-to-reach-62-82-billion-by-2025---increasing-penetration-of-iot--related-devices-drives-market-growth-300974593.html
2. Smith, C.: *Future of AR smartglasses: how they will become the way we view the world* (2019). https://www.wareable.com/ar/future-of-ar-smartglasses-7677
3. Sawh, M.: *Best smart wireless earbuds and hearables to buy 2020* (2019). https://www.wareable.com/hearables/best-hearables-smart-wireless-earbuds
4. Njau, J.: *Smart Shoes: 10 of the Smartest Shoes You Can Put on Your Feet* (2017). https://www.techinsights.com/blog/google-glass-teardown
5. Sawh, M.: *The best smart clothing: from biometric shirts to contactless payment jackets* (2018). https://www.wareable.com/smart-clothing/best-smart-clothing

6. Powell, A.: *Feeling Woozy? Time to Check the Tattoo* (2017). https://news.harvard.edu/gazette/story/2017/09/harvard-researchers-help-develop-smart-tattoos/

7. Dey, N., et al.: Wearable and Implantable Medical Devices: Applications and Challenges. Advances in Ubiquitous Sensing Applications for Healthcare Series. Elsevier Science & Technology (2019). ISBN: 978-0-12-815369-7

8. Moore, G.E.: No exponential is forever: but "forever" can be delayed! In: 2003 IEEE International Solid-State Circuits Conference, 2003. Digest of Technical Papers. ISSCC, vol. 1, pp. 20–23 (2003). https://doi.org/10.1109/ISSCC.2003.1234194

9. Stauth, J.T.: Pathways to mm-scale DCDC converters: trends, opportunities, and limitations. In: 2018 IEEE Custom Integrated Circuits Conference (CICC), pp. 1–8 (2018). https://doi.org/10.1109/CICC.2018.8357017

10. IFIXIT: AirPods Teardown (2016). https://de.ifixit.com/Teardown/AirPods+Teardown/75578

11. IFIXIT: *iPhone 11 Pro Max Teardown* (2019). https://de.ifixit.com/Teardown/iPhone+11+Pro+Max+Teardown/126000

12. IFIXIT: Motorola Moto 360 Teardown (2014). https://de.ifixit.com/Teardown/Motorola+Moto+360+Teardown/28891

13. Foley, R., et al.: Technology roadmapping for power supply in package (PSiP) and power supply on chip (PwrSoC). In: 2010 Twenty-Fifth Annual IEEE Applied Power Electronics Conference and Exposition (APEC), pp. 525–532 (2010). https://doi.org/10.1109/APEC.2010.5433622

14. Steyaert, M., et al.: DCDC converters: from discrete towards fully integrated CMOS. In: 2011 Proceedings of the ESSCIRC, pp. 42–49 (2011). https://doi.org/10.1109/ESSCIRC.2011.6044912

15. Krishnamurthy, H.K., et al.: A digitally controlled fully integrated voltage regulator with on-die solenoid inductor with planar magnetic core in 14-nm tri-gate CMOS. IEEE J. Solid-State Circuits 53(1), 8–19 (2018a). ISSN: 0018-9200. https://doi.org/10.1109/JSSC.2017.2759117

16. Kudva, S.S., Harjani, R.: Fully-integrated on-chip DCDC converter with a 450× output range. IEEE J. Solid-State Circuits 46(8), 1940–1951 (2011). ISSN: 0018-9200. https://doi.org/10.1109/JSSC.2011.2157253

17. Krishnamurthy, H.K., et al.: A 500 MHz, 68% efficient, fully on-die digitally controlled buck voltage regulator on 22nm tri-gate CMOS. In: 2014 Symposium on VLSI Circuits Digest of Technical Papers, pp. 1–2 (2014). https://doi.org/10.1109/VLSIC.2014.6858438

18. Krishnamurthy, H.K., et al.: A digitally controlled fully integrated voltage regulator with 3-D-TSV-based on-die solenoid inductor with a planar magnetic core for 3-D-stacked die applications in 14-nm tri-gate CMOS. IEEE J. Solid-State Circuits 53(4), 1038–1048 (2018b). ISSN: 1558-173X. https://doi.org/10.1109/JSSC.2017.2773637

19. Wens, M., Steyaert, M.S.J.: A fully integrated CMOS 800-mW four-phase semiconstant ON/OFF-time step-down converter. IEEE Trans. Power Electron. 26(2), 326–333 (2011). ISSN: 0885-8993. https://doi.org/10.1109/TPEL.2010.2057445

20. Wens, M., Steyaert, M.: A fully-integrated 0.18 μm CMOS DCDC step-down converter, using a bondwire spiral inductor. In: 2008 IEEE Custom Integrated Circuits Conference, pp. 17–20 (2008). https://doi.org/10.1109/CICC.2008.4672009

21. Lee, M., Choi, Y., Kim, J.: A 0.76 W/mm2 on-chip fully-integrated buck converter with negatively-coupled, stacked-LC filter in 65 nm CMOS. In: 2014 IEEE Energy Conversion Congress and Exposition (ECCE), pp. 2208–2212 (2014). https://doi.org/10.1109/ECCE.2014.6953697

22. Bang, S., Blaauw, D., Sylvester, D.: A successive-approximation switched-capacitor DC–DC converter with resolution of $V_{IN}/2^N$ for a wide range of input and output voltages. IEEE J. Solid-State Circuits 51(2), 543–556 (2016). https://doi.org/10.1109/JSSC.2015.2501985

23. Nguyen, B., et al.: High-efficiency fully integrated switched-capacitor voltage regulator for battery-connected applications in low-breakdown process technologies. IEEE Trans. Power Electron. 33(8), 6858–6868 (2018). https://doi.org/10.1109/TPEL.2017.2757950

24. Salem, L.G., Mercier, P.P.: An 85%-efficiency fully integrated 15-ratio recursive switched-capacitor DCDC converter with 0.1-to-2.2 V output voltage range. In: 2014 IEEE International Solid-State Circuits Conference Digest of Technical Papers (ISSCC), pp. 88–89 (2014). https://doi.org/10.1109/ISSCC.2014.6757350

25. Le, H., Sanders, S.R., Alon, E.: Design techniques for fully integrated switched-capacitor DCDC converters. IEEE J. Solid-State Circuits **46**(9), 2120–2131 (2011). ISSN: 1558-173X. https://doi.org/10.1109/JSSC.2011.2159054

26. Jiang, J., et al.: A 2-/3-phase fully integrated switched-capacitor DCDC converter in bulk CMOS for energy-efficient digital circuits with 14% efficiency improvement. In: 2015 IEEE International Solid-State Circuits Conference – (ISSCC) Digest of Technical Papers, pp. 1–3 (2015). https://doi.org/10.1109/ISSCC.2015.7063078

27. Piqué, G.V.: A 41-phase switched-capacitor power converter with 3.8mV output ripple and 81% efficiency in baseline 90 nm CMOS. In: 2012 IEEE International Solid-State Circuits Conference, pp. 98–100 (2012). https://doi.org/10.1109/ISSCC.2012.6176892

28. Sarafianos, A., Steyaert, M.: Fully integrated wide input voltage range capacitive DCDC converters: the folding Dickson converter. IEEE J. Solid-State Circuits **50**(7), 1560–1570 (2015). ISSN: 1558-173X. https://doi.org/10.1109/JSSC.2015.2410800

29. Hua, X., Harjani, R.: 3.5–0.5 V input, 1.0 V output multi-mode power transformer for a supercapacitor power source with a peak efficiency of 70.4%. In: 2015 IEEE Custom Integrated Circuits Conference (CICC), pp. 1–4 (2015). https://doi.org/10.1109/CICC.2015.7338390

30. Lutz, D., Renz, P., Wicht, B.: A 10 mW fully integrated 2-to-13 V-input buck-boost SC converter with 81.5% peak efficiency. In: 2016 IEEE International Solid-State Circuits Conference (ISSCC), pp. 224–225 (2016a). https://doi.org/10.1109/ISSCC.2016.7417988

31. Lutz, D., Renz, P., Wicht, B.: A 120/230 Vrms-to-3.3 V micro power supply with a fully integrated 17V SC DCDC converter. In: ESSCIRC Conference 2016: 42nd European Solid-State Circuits Conference, pp. 449–452 (2016b). https://doi.org/10.1109/ESSCIRC.2016.7598338

32. Lutz, D., Renz, P., Wicht, B.: An integrated 3-mW 120/230-V AC mains micropower supply. IEEE J. Emerg. Sel. Top. Power Electron. **6**(2), 581–591 (2018). ISSN: 2168-6777. https://doi.org/10.1109/JESTPE.2018.2798504

33. Liu, W., et al.: A 94.2%-peak-efficiency 1.53A direct-battery-hook-up hybrid Dickson switched-capacitor DCDC converter with wide continuous conversion ratio in 65 nm CMOS. In: 2017 IEEE International Solid-State Circuits Conference (ISSCC), pp. 182–183 (2017). https://doi.org/10.1109/ISSCC.2017.7870321

34. Kim, W., Brooks, D., Wei, G.: A fully-integrated 3-LevelDCDC converter for nanosecond-scale DVFS. IEEE J. Solid-State Circuits **47**(1), 206–219 (2012). ISSN: 0018-9200. https://doi.org/10.1109/JSSC.2011.2169309

35. Hardy, C., Le, H.: A 10.9 W 93.4%-efficient (27 W 97%-efficient) flying-inductor hybrid DCDC converter suitable for 1-cell (2-cell) battery charging applications. In: 2019 IEEE International Solid-State Circuits Conference – (ISSCC), pp. 150–152 (2019). https://doi.org/10.1109/ISSCC.2019.8662432

36. Schaef, C., Stauth, J.T.: A 3-phase resonant switched capacitor converter delivering 7.7 W at 85% efficiency using 1.1nHPCBTrace inductors. IEEE J. Solid-State Circuits **50**(12), 2861–2869 (2015). ISSN: 1558-173X. https://doi.org/10.1109/JSSC.2015.2462351

37. Schaef, C., Kesarwani, K., Stauth, J.T.: A variable-conversion-ratio 3-phase resonant switched capacitor converter with 85% efficiency at 0.91 W/mm2 using 1.1nH PCB-trace inductors. In: 2016 IEEE International Solid-State Circuits Conference – (ISSCC) Digest of Technical Papers, pp. 1–3 (2015). https://doi.org/10.1109/ISSCC.2015.7063075

38. Schaef, C., Din, E., Stauth, J.T.: A digitally controlled 94.8%- peak-efficiency hybrid switched-capacitor converter for bidirectional balancing and impedance-based diagnostics of lithium-ion battery arrays. In: 2017 IEEE International Solid-State Circuits Conference (ISSCC), pp. 180–181 (2017). https://doi.org/10.1109/ISSCC.2017.7870320

39. Kesarwani, K., Stauth, J.T.: The direct-conversion resonant switched capacitor architecture with merged multiphase interleaving: cost and performance comparison. In: 2015 IEEE Applied Power Electronics Conference and Exposition (APEC), pp. 952–959 (2015a). https://doi.org/10.1109/APEC.2015.7104464

40. Kesarwani, K., Stauth, J.T.: Resonant and multi-mode operation of flying capacitor multi-level DCDC converters. In: 2015 IEEE 16th Workshop on Control and Modeling for Power Electronics (COMPEL), pp. 1–8 (2015b). https://doi.org/10.1109/COMPEL.2015.7236511

41. Kesarwani, K., Sangwan, R., Stauth, J.T.: A 2-phase resonant switched-capacitor converter delivering 4.3 W at 0.6 W/mm2 with 85% efficiency. In: 2014 IEEE International Solid-State Circuits Conference Digest of Technical Papers (ISSCC), pp. 86–87 (2014). https://doi.org/10.1109/ISSCC.2014.6757349

Chapter 2
Motivation and Fundamentals

2.1 Power Delivery in Portable Systems

There are significant differences in the hardware specification between handheld devices (e.g., smartphones and tablets) and wearables (e.g., smartwatches) [1–4]. The battery capacity of a smartwatch (200–400 mAh) is typically an order of magnitude smaller than a smartphone battery with a typical value of 2000 mAh. For long battery runtime, the display of the smartwatch is smaller and equipped with fewer pixels. The CPU is a scaled-down version, which operates at lower frequencies. Table 2.1 gives an overview of the power consumption of a typical smartwatch [1–4]. Different operation modes (sleeping, awake) and the additional power consumption of other components (CPU, Bluetooth, Wi-Fi, and display) are listed. In sleep mode, the smartwatch has a low power consumption of <25 mW. In active operating mode, the power consumption can rise up to several hundreds of milliwatts through display interaction and the transmission of information via Bluetooth and Wi-Fi.

The loss breakdown of a smartwatch in Fig. 2.1 indicates that more than half of the energy is spent when the watch is in sleep mode [3–5]. The active periods of watches are much shorter than those of phones due to the nature of the applications in watches, e.g., time checking, notifications, etc. In active mode, the display is still the dominant power consumer with 17.7%. Bluetooth and Wi-Fi play a less important role in power consumption due to the low volume data traffic generated by smartwatches.

Most portable electronic systems are powered by a battery. Table 2.2 shows an overview of typical characteristics of commonly used rechargeable batteries in portable systems [6, 7]. Li-ion batteries are the most popular choice for portable devices due to their superior energy density, which helps in minimizing the size and weight. The self-discharge of Li-ion batteries is also less than half of NiCd and NiMH batteries. During its discharge cycle, the Li-ion battery provides an output

© The Editor(s) (if applicable) and The Author(s), under exclusive license to
Springer Nature Switzerland AG 2021
P. Renz, B. Wicht, *Integrated Hybrid Resonant DCDC Converters*,
https://doi.org/10.1007/978-3-030-63944-0_2

Table 2.1 Power consumption of a typical smartwatch (LG Watch Urbane) [3, 4]

Modes/component	Power consumption/mW
Sleep mode (watch face off)	10.9
Sleep mode (watch face on)	23.5
Awake (basic consumption)	43.5
CPU	0.6
Bluetooth	180
Wi-Fi	400
Display (touch/swipe)	120

Fig. 2.1 Loss breakdown of a smartwatch from a user study [3, 4]

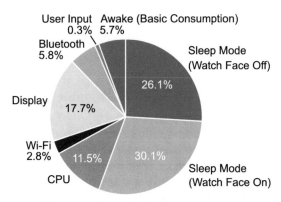

Table 2.2 Characteristics of commonly used rechargeable batteries in portable electronic systems

	NiCd	NiMH	Reusable alkaline	Li-ion	Li-ion polymere
Energy density/Wh/kg	45–80	60–120	80	110–160	100–130
Voltage/V	1.2–1.6	1.0–1.5	0.9–1.5	3.0–4.5	3.0–4.5
Cycle life	1500	300–500	50	500–1000	300–500
Self-discharge/month	20%	30%	0.3%	10%	10%
Costs	Low	High	Medium	High	High

voltage, which can widely vary from 4.5 to 3.0 V with a nominal cell voltage of around 3.6 to 3.7 V. Since this is different from the nominal supply voltage of the application chip in a given CMOS technology ($\sim 1.0 - 1.8$ V), a power management unit with a DCDC converter is required.

Figure 2.2a shows a typical implementation of a battery-operated wearable or IoT application. A discrete power management IC (PMIC) is used together with external passive components for powering the application chip. Often inductive switching topologies are used because of their good efficiency and controllability. However, the employed passive components are large, which increase PCB area and costs.

To meet the needs of the future generations of wearable and IoT devices, the goal is to integrate all required power conversion components on one chip (SoC) as shown in Fig. 2.2b. To achieve this, the size of the passive components must be

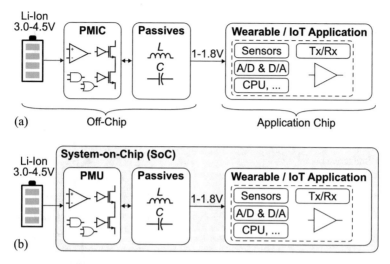

Fig. 2.2 Block diagram of battery-operated wearable or IoT application: (**a**) conventional powering with a PMIC together with external passive energy storage components; (**b**) fully integrated power management implemented as a SoC

drastically minimized. Therefore, higher switching frequencies and new conversion approaches and techniques are necessary. The key challenge is to implement a power- and area-efficient solution.

There are many different options for implementing a fully integrated DCDC converter. A short overview of these conversion principles and an assessment of the capability for efficient fully integrated voltage conversion are given in the following section.

2.2 Types of DCDC Converters

A DCDC conversion can be performed by a resistive regulator (linear regulator), capacitively (switched-capacitor converter), inductively (inductive converter), or with a combination of inductive and capacitive concepts (hybrid converter). This section gives a short overview of different DCDC converter types.

2.2.1 Linear Regulators

Due to their simplicity, linear regulators are widely used in complex ICs, especially in power management units (PMU). Figure 2.3 shows the basic principle. The output voltage results in $V_{\text{out}} = V_{\text{in}} - R \cdot I_{\text{out}}$. Depending on the output current I_{out}, the

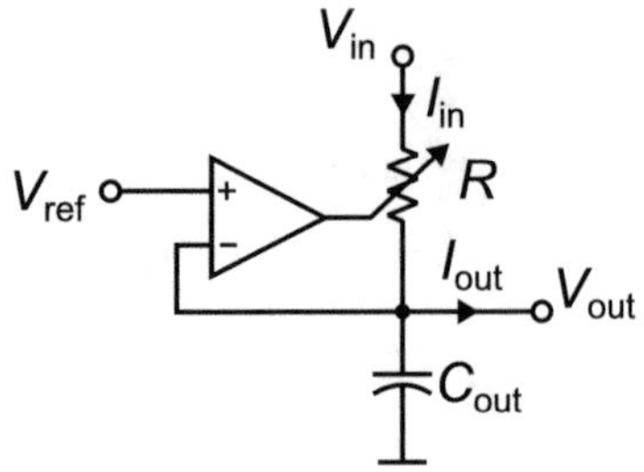

Fig. 2.3 Linear regulator principle

Fig. 2.4 Inductive buck converter

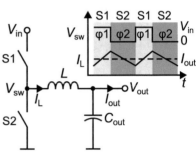

resistance value of the resistor R is adjusted to ensure a constant output voltage as indicated in Fig. 2.3. The resistor R is usually implemented as a regulated NMOS or PMOS transistor. In the ideal case, the input current I_{in} equals the output current I_{out}.

The efficiency η_{LR} of an ideal linear regulator can be expressed by

$$\eta_{LR} = \frac{P_{out}}{P_{in}} = \frac{V_{out} \cdot I_{out}}{V_{in} \cdot I_{in}} = \frac{V_{out}}{V_{in}}. \tag{2.1}$$

The major drawback of linear regulators follows from Eq. 2.1. As the power losses increase with rising input voltage V_{in} and falling output voltage V_{out}, very low efficiency η_{LR} results for high conversion ratios $VCR = V_{in}/V_{out}$. Thus, linear voltage regulators are mostly used in applications with a small voltage difference between the input and output or where a low efficiency can be accepted.

2.2.2   Inductive Converters

Inductive converters use switches and inductors to transfer energy from the input to the output. The principle of a simple and widely used inductive buck converter is shown in Fig. 2.4. It is also referred to as two-level buck converter since it provides two different voltage levels at the switching node V_{sw}, either V_{in} or ground. A pulse-width modulated (PWM) signal is generated at the switching node V_{sw} by means of the switches S1 and S2, which is then filtered through an L-C filter. Different DC voltage levels can be generated by varying the duty cycle of the PWM signal.

Inductive converters lead to significantly higher efficiency compared to linear regulators. Assuming ideal devices and passives, a theoretical efficiency of 100% can be achieved independent of the voltage conversion ratio VCR (input and output voltage). Non-ideal components degrade the efficiency in a real implementation, for instance, the on-resistance of the switches, parasitic capacitances of the switches, and inductor losses (ESR, core losses). Nevertheless, inductive buck converters can achieve very high efficiencies >90% [8–11]. The typical switching frequency is relatively low, which requires off-chip filter components since the resonance frequency of the L-C filter has to be much lower than the switching frequency for proper PWM operation. On-chip integration of the filter components requires very high switching frequencies (>100 MHz) since only small inductance and capacitance values can be integrated. This increases the switching losses in both active and passive components significantly. Together with the higher conduction losses, due to low-quality factor of on-chip inductors, the efficiency is degraded significantly [12–18]. Only small voltage conversion ratios VCR are obtained due to the large voltage drop across the inductor and the power switches, which leads to large inductor current ripple and higher switching losses.

Hybrid converters, introduced in Sect. 2.2.4, can overcome these challenges by introducing additional capacitors, which generate intermediate- and lower-voltage levels.

2.2.3 Switched-Capacitor Converters

In switched-capacitor (SC) DCDC converters, switches and capacitors are used to perform a voltage conversion. Capacitors are responsible for the charge transfer from the input to the output of the converter. Figure 2.5 shows a series-parallel converter, which divides the input voltage V_{in} by a factor of two. In the first phase $\varphi 1$, the flying capacitor C_{fly} is connected in series with the output capacitor C_{out} and is charged from the input voltage V_{in}. In the second phase $\varphi 2$, the flying capacitor is connected between ground and the output capacitor. Phase $\varphi 1$ is the charging phase, while phase $\varphi 2$ is the discharging phase with respect to the flying capacitor. In steady

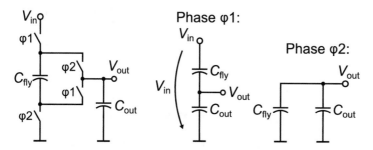

Fig. 2.5 Principle of a series-parallel SC converter

state, the flying capacitor has the same voltage as V_{out} due to charge balancing. If the output voltage and the input voltage are swapped, the converter becomes a boost structure, which doubles the voltage. The major advantage of SC converters is their capability for monolithic integration since capacitors have a higher energy density than inductors and can be integrated with a higher-quality factor in standard CMOS processes [19, 20].

The voltage conversion ratio N is determined by the circuit topology and not by the duty cycle as in inductive converters. High efficiencies can be achieved close to the ideal conversion ratio of the converter. In case of a load current, charge must be transferred from the input to the output, and the capacitors must be charged and discharged, causing a voltage drop at the output of the converter. This voltage drop is proportional to the output current, which can be modeled by an equivalent output resistance R_{out}, introduced in Sect. 2.3.2. Due to the capacitive charge transfer losses, the efficiency η_{SC} is reduced similar to a linear voltage regulator as described by

$$\eta_{SC} = \frac{V_{out}}{V_{in} \cdot N}. \tag{2.2}$$

Multi-ratio architectures, which realize several conversion ratios N, can be used to increase the efficiency over a wide input voltage range. Figure 2.6 shows the theoretical efficiency η (not covering switching and control losses) of a switched-capacitor converter versus the input voltage V_{in} for varying conversion ratios N. The corresponding ratio is chosen depending on the operating point. A higher number of ratios N leads to a significant improvement of the efficiency minima. But it also leads to increasing complexity and chip area, since more switches and flying capacitors are required for topology reconfiguration. Higher input voltages also require additional supporting circuits like level shifters and a generation of

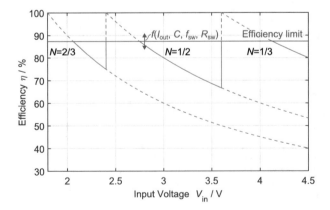

Fig. 2.6 Theoretical efficiency η of a switched-capacitor converter versus the input voltage V_{in} for varying conversion ratios N

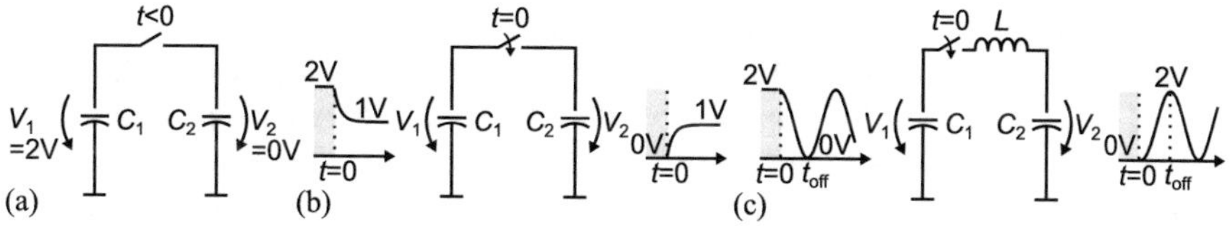

Fig. 2.7 Charge sharing circuit of two capacitors C_1 and C_2: (**a**) disconnected; (**b**) shorted; (**c**) connected via an inductor L

auxiliary supply rails for the control of the power switches [21–25]. Indicated in the picture, the theoretical efficiency is reduced by capacitive charge transfer losses, resistive losses, and control losses. The efficiency limit also depends on the design of the capacitors C and the switching frequency f_{sw}. Especially the capacitive charge transfer losses limit the maximum achievable output current and output power. This is why SC converters are often only used in low-power applications [21–23, 26–32].

Since the capacitive charge transfer losses have been identified as the main bottleneck in SC converters, they are analyzed for a simple example in the following. For more complex converter structures, the charge transfer losses can be calculated using the equivalent output resistance approach, introduced in Sect. 2.3.2.

Assuming, there are two identical capacitors $C_1 = C_2 = 1\,F$, one charged to $V_1 = 2\,V$ and the other one fully discharged to $V_2 = 0\,V$ with a switch between them as shown in Fig. 2.7a, the total energy stored in the capacitors can be calculated by

$$E_{tot} = \frac{1}{2} \cdot C_1 \cdot V_1^2 + \frac{1}{2} \cdot C_2 \cdot V_2^2 \tag{2.3}$$

For Fig. 2.7a, this results in $E_{tot,(a)} = 2\,J$. If the switch is closed at $t = 0$, the capacitors are connected in parallel, as shown in Fig. 2.7b. The total charge will remain the same before and after the switch is closed. However, the equivalent capacitance is $C_1 + C_2 = 2\,F$ after the switch is closed. The voltage across the capacitors will settle to $V_1 = V_2 = 1\,V$, following an exponential curve as indicated in Fig. 2.7b. This leads to a total stored energy of $E_{tot,(b)} = 1\,J$. It can be observed that by connecting the two capacitors and sharing the charge, half of the initial energy is lost. These charge sharing losses occur in all SC converters and can only be reduced by limiting the voltage difference between the capacitors. As the capacitors are not fully charged and discharged in this case, only a small fraction of the available capacitance is used for the energy transfer. This leads to the need of high switching frequencies, especially for fully integrated converters with small capacitance values.

The "missing" energy, which is dissipated as heat in the resistance in the path, can be saved by converting it into a new energy storage form. In addition to the storage in the electric field, an inductor L can be used for storing energy in a magnetic field as shown in Fig. 2.7c. This gives the possibility of oscillation. The current from C_1 to C_2 via the inductor L increases with a sinusoidal shape, as long as V_1 is higher than V_2. At $V_1 = V_2$, the current has reached its maximum. The inductor forces the current flow to continue. V_2 becomes higher than V_1 to a maximum of $V_2 = 2\,V$. At this time, the current as well as the voltage V_1 approaches zero. This is the point where the switch should be opened again, labeled by t_{off} in Fig. 2.7c. The total energy can be calculated by Eq. 2.3, which results in $E_{tot,(c)} = 2\,J$ for the example values introduced above. This means that the whole charge and energy are transferred from C_1 to C_2 and no charge sharing losses occur. In reality there are some losses due to finite conductivity of the switch and the inductor. It is worth noting that, if the switch is not opened again, the stored energy will oscillate back and forth between the capacitors and the inductor until the current reaches zero, storing the remaining energy in the capacitors ($V_1 = V_2 = 1\,V$) like in the pure SC case of Fig. 2.7a, b. Again, during this process, half of the energy is lost to heat in the resistance.

Identical observations can be made for mechanical resonances, e.g., in a mass-spring oscillator where the stored energy changes its form between potential and kinetic energy. The equivalent of kinetic energy is the magnetic energy in the inductor, while the potential energy is the equivalent of the energy stored in the capacitor. The voltage corresponds to the amplitude of the deflection, and the electrical resistance corresponds to the friction.

Based on these investigations, an additional inductor can be used to leverage the potential of SC converters. This leads to hybrid converters described in the following section.

2.2.4 Hybrid Converters

Hybrid converters are a promising converter class, which merge the capacitive and inductive conversion approaches into one. A drastic reduction of the inductor value can be achieved while minimizing switching losses and improving the overall efficiency [33–44]. Depending on the point of view (SC or inductive), a "hybridization" of the converter either eliminates the charge transfer losses (SC converter) or reduces the voltage across the inductor and the switches (inductive converter). Sometimes a simple parallel connection of two converters, one inductive and one capacitive, is called a hybrid converter [45, 46]. For high output power, the inductive converter is used, while for smaller output power, the capacitive converter takes over. Due to the ineffective component utilization and circuit overhead, this approach is not discussed in detail in the following. A cascade connection of a capacitive converter followed by an inductive converter is presented in [47–50]. It eliminates the SC regulation problem and enables smaller inductor values. However, separate power stages increase the resistance in the power path. Additional to that, a relatively bulky intermediate buffer capacitor is needed for balancing of the capacitor cells. In the following, approaches are investigated, which merge the capacitive and inductive converter into one structure such that the components are shared between them and the intermediate capacitor can be eliminated.

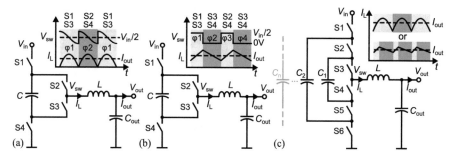

Fig. 2.8 Different hybrid converter examples: (**a**) resonant SC (ReSC) converter; (**b**) three-level buck converter (waveforms shown for $V_{out} \leq V_{in}/2$); (**c**) extension of (**a**) or (**b**) to N-multilevel converter

Figure 2.8 shows examples of different hybrid converter topologies. The resonant switched-capacitor (ReSC) converter in Fig. 2.8a consists of a series-parallel SC stage together with an inductor L (see also Sect. 2.3). It provides a conversion ratio of 1/2 identical to Fig. 2.5. The charge sharing losses of traditional SC converters can be eliminated with an inductor L in series since it decouples the flying capacitor from the output capacitor and enables resonant charging and discharging (see Fig. 2.7 in Sect. 2.2.3) [36–44, 51–53]. In this approach, there is no need to limit the voltage ripple on the flying capacitors as with SC converters. The full available capacitance C_{fly} can be used because it may be even fully discharged as outlined in Fig. 2.7c. In consequence, this topology is very suitable for integrated circuits. Even small inductance values in the range of 1–10 nH are sufficient to ensure resonant operation with integrated caps at reasonable frequencies. Operation at the resonance frequency $f_{sw,res} = 1/(2\pi\sqrt{LC})$ has the additional benefit of zero current switching (ZCS) as depicted in Fig. 2.8a. With ZCS, the power switches are turned on and off exactly when the current is zero, which eliminates the transition losses. With this approach, highly integrated converters with very small external inductors can be realized [37, 38, 44], which achieve high efficiencies of up to 85%.

The power stage of an inductive three-level buck converter is depicted in Fig. 2.8b. It has the same structure as the ReSC converter in Fig. 2.8a, but the control of the switches is different. Two more phases $\varphi 3$ and $\varphi 4$ are introduced, and the switching frequency is significantly higher than the resonance frequency $f_{sw,res}$, required for the duty cycle regulation of the PWM signal. This leads to the sawtooth-based current waveforms for the inductor current as shown in Fig. 2.8b. The three-level buck converter can be seen as an extension of the conventional buck converter shown in Fig. 2.4 by the flying capacitor C. In this configuration, the capacitor C can be considered as a constant voltage source with its voltage balanced at $V_{in}/2$. Three different voltage levels can be provided at the switching node V_{sw}, namely, V_{in}, $V_{in}/2$, and 0 V. When the output voltage is higher than half the input voltage ($V_{out} > V_{in}/2$), the switching node V_{sw} alternates between V_{in} and $V_{in}/2$. Likewise, when the output voltage is lower than half the input voltage ($V_{out} < V_{in}/2$), the switching node alternates between $V_{in}/2$ and 0 V. Compared

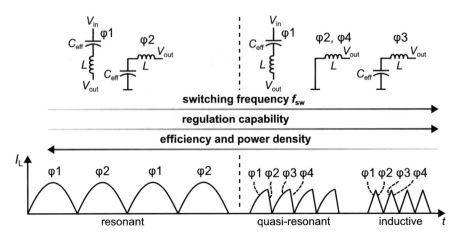

Fig. 2.9 Overview of different operation modes of hybrid converters

to the traditional two-level buck converter (Fig. 2.4), the three-level buck converter ensures a lower voltage across the inductor and therefore achieves a much smaller inductor current ripple. Overall, the 3-level concept enables a significant reduction of the size of the inductor L and output capacitor C_{out} compared to a conventional two-level buck converter. The switches in the three-level converter need to block only half the input voltage, which allows to use low-voltage transistors. This reduces the chip area and also the associated switching loss. Further details can be found in [54]. Compared to ReSC converters, fully integrated three-level buck converters suffer from lower efficiencies [34, 54–57] due to their high switching frequency required for PWM regulation (see Sect. 2.2.5). This is also the limiting factor for efficient low-power operation.

Both converter types (ReSC and inductive) can be further extended by additional flying capacitors for higher conversion ratios VCR. This leads to hybrid SC stages [20, 33, 35, 58–63] in various configurations. Some topologies use more complex SC stages like Dickson and series-parallel. Others lead to flying capacitor multilevel (FCML) converters where N multiple levels can be generated at the switching node as indicated in Fig. 2.8c. FCML converters are mostly treated as inductive multilevel buck converters [64–67], but many more operation modes are possible [68, 69].

Different operation modes are shown in Fig. 2.9 along with their current waveforms. Resonant and quasi-resonant operation is often applied when the output voltage is close to the ideal SC conversion ratio [37, 38, 53, 70]. Quasi-resonant operation is a mixture between resonant and inductive operation as indicated in the central part of Fig. 2.9. Inductive PWM operation is used when the output voltage is far from the ideal SC ratio [43, 65–67]. In that case the switching frequency is significantly higher than the resonance frequency of the L-C filter. In resonant operation, two switching phases $\varphi 1$ and $\varphi 2$ are required where the effective flying capacitors $C_{fly,eff}$ and the inductor L are always connected in series as indicated in the left part of Fig. 2.9. For quasi-resonant and inductive operation, more phases φ_1–

φ_4 are required for regulation. The inductor is connected to ground or to V_{in} during the switching phases φ_2 and φ_4. In principle, a higher switching frequency leads to better regulation capability (less regulation losses) but also to higher switching losses. This limits the maximum achievable efficiency especially in fully integrated converters [34, 54–57, 65], which can only be improved by larger filter components [33, 37, 38, 71]. This contradicts the goal of full integration. Resonant operation at lower switching frequencies leads to high efficiencies but only around the ideal SC conversion ratio of the power stage [38, 44, 62, 63].

2.2.5 State-of-the-Art Converters

A selection of relevant highly integrated hybrid DCDC converters is depicted in Fig. 2.10. The peak efficiency is plotted over the dynamic load current range. The dynamic load current range $DLCR$ is defined as the ratio between the maximum and minimum load current

$$DLCR = \frac{I_{out,max}}{I_{out,min}} \tag{2.4}$$

where the converter still has an efficiency higher than 80% of its peak efficiency. The color of the markers indicates if the converters are capable of handling the high input voltage range of a Li-ion battery. The marker type represents the operation mode (resonant, quasi-resonant, or inductive) of the converters.

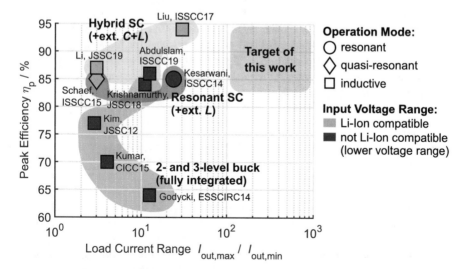

Fig. 2.10 Efficiency versus dynamic load current range of published highly integrated state-of-the-art hybrid DCDC converters [12, 33, 34, 37, 44, 55, 57, 61]

Figure 2.10 shows that fully integrated two- or three-level buck converters suffer from low efficiencies due to the high switching frequency required for inductive operation. They also do not support an effective low-power operation (<1 mA) and operate at low input voltages of $V_{in} < 2.5$ V [12, 34, 55, 57]. With the usage of external capacitors (1–70 µF) and inductors (12–180 nH), inductive PWM-controlled hybrid SC converters can achieve higher efficiencies but also without supporting an effective low-power operation [33, 61].

Resonant SC converters offer very high efficiencies and high power operation while significantly reducing the flying capacitor (18–24 nF) and inductor size (10–17 nH) [37, 44]. However, low-power operation is not supported so far. Highly integrated solutions are possible since the flying capacitors can be integrated on-chip, while the inductors are formed with PCB-integrated trace inductors [37] or directly placed on top of the die [44].

The goal of this work is to cover the full Li-ion battery range with high conversion efficiency required for a long battery runtime. The converter has to be capable to cover a wide output current range from high power to low power since wearables and portable applications are often in sleep or idle mode (see Sect. 2.1). Since most of these applications are small and have limited space, a small form factor with the potential for full integration is an additional condition for the DCDC converter.

As it shows the highest potential for full integration with high conversion ratio, the focus of this work is on the resonant SC converter. The challenge is to find an architecture along with a suitable control technique, which can efficiently cover a wide input voltage and output current range (see Chap. 3). A performance optimized low-power design of all implemented circuit blocks is crucial to achieve high efficiency while switching at high frequencies (see Chap. 4). This also applies to the design of on-chip inductors, which requires a high-quality factor for resonant operation. Implementation and design details are discussed in Chap. 5.

A multi-ratio approach is chosen since it enables efficient resonant operation at moderate resonance frequencies instead of quasi-resonant or inductive regulation where the converter has to operate at significantly higher switching frequencies degrading the overall efficiency. Obviously, a multi-ratio power stage increases the circuit complexity, but this challenge is well addressed by full integration.

2.3 Fundamentals of Resonant Switched-Capacitor Converters

In this section, the fundamentals and the relationships of the converter parameters are introduced. Sections 2.3.1 and 2.3.2 give an overview on the charge flow analysis and the equivalent output resistance model of resonant SC converters. It comprises the intrinsic losses (capacitive charge transfer losses and conduction losses) and also models the steady-state behavior of the converter. The equivalent output resistance is an important part of the efficiency model (see Sect. 3.3) as well as of the dynamic model (see Sect. 6.2). The extrinsic losses introduced by the implementation of the converter are calculated in Sect. 2.3.3.

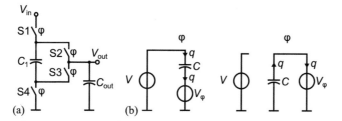

Fig. 2.11 (a) Topology of a 1/2 series-parallel converter; (b) charge flow analysis for corresponding phase $\varphi 1$ and $\varphi 2$

2.3.1 Charge Flow Analysis

The charge flow analysis, introduced in [72], is a good tool for a detailed analysis of the functionality and internal processes even in more complex SC topologies. It can be also used for analyzing resonant SC converters (see Sect. 2.3.2). This analysis is based on the work in [73, 74], which develops a fundamental model of SC converters and introduces the concept of the slow-switching limit (SSL) impedance. In the SSL approximation, the parasitic resistance is neglected. The matrix-based methods of [73, 74] are significantly simplified and generalized by considering the fast-switching limit (FSL) impedance where the parasitic resistance is dominant. With the charge flow analysis, the conversion ratio N and the equivalent output resistance R_{out} (see Sect. 2.3.2) of a converter can be calculated. Additional to that, it can be used for properly sizing of the capacitors and switches. In Sects. 2.3.2 and 3.3, it is also used for analyzing resonant SC converters.

A first step in the charge flow analysis is to identify the different capacitor configurations for the separate phases $\varphi 1$ and $\varphi 2$ of the conversion as shown in Fig. 2.11b for a simple series-parallel SC converter with $N = 1/2$ (Fig. 2.11a). For each phase i, a charge multiplier vector can be defined describing the topology based on the charge flow through the capacitors.

$$\boldsymbol{a}^{(i)} = \left[q_{\text{out}}^{(i)} \; q_{C_1}^{(i)} \; \cdots \; q_{C_n}^{(i)} \; q_{\text{in}}^{(i)} \right]^{\text{T}} \cdot \frac{1}{q_{\text{out}}} = \left[a_{\text{out}}^{(i)} \; a_{C_1}^{(i)} \; \cdots \; a_{C_n}^{(i)} \; a_{\text{in}}^{(i)} \right]^{\text{T}} \quad (2.5)$$

Each element of a charge vector $a^{(i)}$ corresponds to a specific capacitor (e.g., $q_{C_1}^{(i)}$) or voltage source ($q_{\text{out}}^{(i)}$ and $q_{\text{in}}^{(i)}$) and represents the charge flow into that component, normalized with respect to the total output charge flow q_{out} (e.g., $q_{\text{out}} = q_{\text{out}}^{(1)} + q_{\text{out}}^{(2)}$, for $i = 2$). These charge vector elements can be determined by inspection for every state of the conversion period based on the following principles [73]. The sum of the charge flow elements equals zero in each circuit node (Kirchhoff's Current Law). The charge flow in each phase equals zero in steady state. The output and input voltage is represented by a constant voltage source. For the series-parallel converter in Fig. 2.11a, the charge flow analysis results in (with $q_{\text{out}} = 2q_{\text{x}}$)

$$\boldsymbol{a}^{(1)} = [q_x \ q_x \ q_x]^T \cdot \frac{1}{2q_x} = \left[\frac{1}{2} \ \frac{1}{2} \ \frac{1}{2}\right]^T \tag{2.6}$$

$$\boldsymbol{a}^{(2)} = [q_x \ -q_x \ q_x]^T \cdot \frac{1}{2q_x} = \left[\frac{1}{2} \ -\frac{1}{2} \ 0\right]^T . \tag{2.7}$$

Equations 2.6 and 2.7 indicate that the coefficients of the two output components sum to one while the charge vector elements of the capacitor have opposite signs. The ratio of the total input and output charge elements defines the conversion ratio N of the topology [72]

$$N = \frac{\sum a_{in}^{(i)}}{\sum a_{out}^{(i)}} \tag{2.8}$$

For the series-parallel converter in Fig. 2.11a, a conversion ratio $N = 1/2$ is obtained. The charge flow analysis is an easy-to-use method to determine the conversion ratio especially for more complex SC topologies. In Sect. 2.3.2 it is used for the derivation of the equivalent output resistance R_{out}.

2.3.2 Equivalent Output Resistance

The equivalent output resistance model shown in Fig. 2.12 is a common method for modeling the behavior of SC and ReSC converter. An ideal transformer represents the ideal conversion ratio N. The output impedance R_{out} of the converter represents all intrinsic losses, which include the capacitive charge transfer losses as well as the conduction losses. With this model, the load behavior of the converter can be modeled on one hand, and, on the other hand, it is suitable for efficiency analyses and optimization (see Sect. 3.3). For an accurate efficiency model, the frequency-dependent losses such as gate charging losses, parasitic bottom-plate capacitor losses, and losses in level shifters, charge pumps, etc. have to be considered additionally. This is described in Sect. 2.3.3.

The equivalent output resistance model was introduced in [72–75] for SC converters. In [76–80] it was then extended for ReSC converters. The derivation of the output impedance R_{out} in [80] is not valid in general since it is derived for a simplified resonant case where an ideal blocking diode prevents negative currents.

Fig. 2.12 Generalized equivalent output resistance model for SC converter and ReSC converter

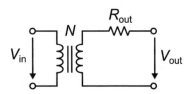

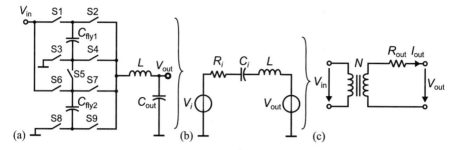

Fig. 2.13 Derivation of equivalent output resistance: (**a**) schematic of a multi-ratio resonant SC power stage; (**b**) equivalent circuit for any phase i; (**c**) generalized equivalent output resistance model

A generalized output impedance is derived in [76–79] for a fixed 1/2 conversion ratio where several simplifications can be applied. References [78, 79] use the charge multiplier framework from [72], which is already well-known from SC converter analysis.

The calculations in this work are based on [78, 79] and extended for further conversion ratios and different duty cycles (Sect. 3.1). As an example, a multi-ratio resonant SC power stage is shown in Fig. 2.13a. The structure corresponds to the actual used power stage, which is introduced in Sect. 3.1. It is assumed that any single phase i of a ReSC converter depicted in Fig. 2.13a can be modeled by an equivalent circuit consisting of an capacitor C_i, an resistor R_i and an inductor L as shown in Fig. 2.13b. The components C_i, R_i and L can be derived from complex networks of multiple elements. The internal voltage V_i and the output voltage V_{out} drive the effective voltage across the capacitor (initial conditions on C_i are lumped into V_i). The period of the individual switching phases is determined by the switching frequency and the respective duty cycles. Each of the switching phases, which are modeled by the corresponding equivalent circuits, is responsible for a power loss, which is dissipated by the total subcircuit resistance R_i. This power loss is referenced to the output current I_{out}, which serves as a common reference. The output current can be calculated by

$$I_{\text{out}} = f_{\text{sw}} \cdot \sum q_{\text{out}}^{(i)} \tag{2.9}$$

where $q_{\text{out}}^{(i)}$ is the charge transferred to the output in each phase i, which is known from the charge flow analysis in Sect. 2.3.1. With the definition of the equivalent output resistance, the average power dissipated in the converter can be calculated from

$$P = f_{\text{sw}} \cdot \sum E_i = I_{\text{out}}^2 \cdot R_{\text{out}} = \left(f_{\text{sw}} \sum q_{\text{out}}^{(i)} \right)^2 R_{\text{out}}. \tag{2.10}$$

The energy E_i dissipated during phase i can be defined by

$$E_i = \int_0^{\tau_i} P_i(t)\mathrm{d}t = \int_0^{\tau_i} R_i \cdot I_i{}^2(t)\mathrm{d}t. \tag{2.11}$$

Equating 2.10 and 2.11 and solving for the equivalent output resistance lead to

$$R_{\text{out}} = \frac{\sum \left(\int_0^{\tau_i} R_i \cdot I_i{}^2(t)\mathrm{d}t \right)}{f_{\text{sw}} \cdot \left(\sum q_{\text{out}}^{(i)} \right)^2}. \tag{2.12}$$

The summations are proceeded over all phases i. According to the charge flow analysis, $\sum q_{\text{out}}^{(i)}$ can be expressed with

$$\sum q_{\text{out}}^{(i)} = \sum a_{\text{out}}^{(i)} \cdot q_{\text{out}} \tag{2.13}$$

and Eq. 2.12 can be rewritten to

$$R_{\text{out}} = \frac{\sum \left(\int_0^{\tau_i} R_i \cdot I_i{}^2(t)\mathrm{d}t \right)}{f_{\text{sw}} \cdot \left(\sum \left(a_{\text{out}}^{(i)} \cdot q_{\text{out}} \right) \right)^2}. \tag{2.14}$$

Equation 2.14 gives a generalized expression for the equivalent output resistance of SC or ReSC converters. It is a function of the frequency f_{sw}; the charge vector element $a_{\text{out}}^{(i)}$, which depends on the conversion ratio; the resistance R_i; and the corresponding current $I_i(t)$ in the phase i. For a ReSC converter, the current $I_i(t)$ can be calculated with the equivalent circuit in Fig. 2.13b. The same circuit can be used for the calculation of the current $I_i(t)$ for an SC converter by simply neglecting the inductance L. The equivalent output resistance can then be used for efficiency calculation and optimization (see Sect. 3.3) as well as for the dynamic model of the ReSC converter (see Sect. 6.2).

Current Equation for a ReSC Converter

For the calculation of the equivalent output resistance with Eq. 2.14, the current $I_i(t)$ has to be calculated for the ReSC converter. The equivalent circuit in Fig. 2.13b leads to a homogeneous second-order differential equation with constant coefficients.

$$0 = L C_i \cdot \frac{\mathrm{d}^2 I_i}{\mathrm{d}t^2} + R_i C_i \cdot \frac{\mathrm{d}I_i}{\mathrm{d}t} + I_i \tag{2.15}$$

This can be solved to get the generalized solution for the current $I_i(t)$ in each phase i

$$I_i(t) = A_{i1} \cdot e^{\lambda_{i1}t} + A_{i2} \cdot e^{\lambda_{i2}t} \tag{2.16}$$

For $I(t)$ this results in

$$I(t) = \begin{cases} A_{11} \cdot e^{\lambda_{11}t} + A_{12} \cdot e^{\lambda_{12}t} & t \in [0, \tau_1] \\ A_{21} \cdot e^{\lambda_{21}(t-\tau_1)} + A_{22} \cdot e^{\lambda_{22}(t-\tau_1)} & t \in [\tau_1, \tau_1 + \tau_2] \end{cases} \tag{2.17}$$

where the eigenvalues λ_{ik} are

$$\lambda_{ik} = \frac{-R_i}{2L} \pm \sqrt{\left(\frac{R_i}{2L}\right)^2 - \frac{1}{LC_i}}. \tag{2.18}$$

For the determination of the constants A_{ik}, four additional boundary conditions are required. The first two can be determined by means of the charge flow analysis. The charge $q_{out}^{(i)}$ flowing into the output in each phase i must be equal to the integral of the current over the phase τ_i

$$q_{out}^{(i)} = a_{out}^{(i)} \cdot q_{out} = \int_0^{\tau_i} I_i(t)dt \tag{2.19}$$

The other two boundary conditions can be determined by the transition during the phases. Due to the inductor, the current at the beginning of each switching phase must be the same as at the end of the previous one. For two switching phases ($i = 2$), this means

$$I_1(t = 0) = I_2(t = \tau_2) \tag{2.20}$$

$$I_2(t = 0) = I_1(t = \tau_1) \tag{2.21}$$

Inserting Eq. 2.17 in Eqs. 2.19, 2.20, and 2.21 leads to a four-by-four equation system for the determination of A_{11} to A_{22}. Equation 2.17 can be inserted in Eq. 2.14 in order to calculate the equivalent output resistance of a ReSC converter. Several simplifications can be made for the conversion ratio $N = 1/2$, e.g., $C_1 = C_2$, $R_1 = R_2$, and $\tau_1 = \tau_2$, which significantly simplify the calculations [76–79]. However, for the general case with arbitrary conversion ratios, these assumptions no longer apply. Different capacitor configurations ($C_1 \neq C_2$, $R_1 \neq R_2$) lead to varying resonance frequencies and duty cycles ($\tau_1 \neq \tau_2$) in both switching phases, as described in Sect. 3.1. Therefore, the symbolic solution of the four-by-four equation system and for R_{out} leads to a very complex equation. For this reason, MATLAB® is used for solving the system of equations and the final R_{out} equation. With this it is possible to determine the equivalent output resistance for different conversion ratios and duty cycles.

Current Equation and Equivalent Output Resistance for a SC Converter

At higher switch resistances (see Sect. 3.2), the influence of the inductor L can be neglected, and the converter can be treated like a conventional SC converter. This eliminates the inductor L in the equivalent circuit in Fig. 2.13 and leads to a homogeneous first-order differential equation with constant coefficients for the current $I_i(t)$.

$$I_i(t) = A_i \cdot e^{-t/(R_i C_i)} \tag{2.22}$$

The constant A_i can be determined with the following boundary condition. The charge $q_{\text{out}}^{(i)}$ flowing into the output in each phase i must be equal to the integral of the current over the phase τ_i

$$q_{\text{out}}^{(i)} = a_{\text{out}}^{(i)} \cdot q_{\text{out}} = \int_0^{\tau_i} I_i(t)\mathrm{d}t = \int_0^{\tau_i} A_i \cdot e^{-t/(R_i C_i)}\mathrm{d}t. \tag{2.23}$$

Solving Eq. 2.23 for A_i leads to

$$A_i = \frac{a_{\text{out}}^{(i)} \cdot q_{\text{out}}}{R_i C_i (1 - e^{-\tau_i/(R_i C_i)})}. \tag{2.24}$$

The equivalent output resistance R_{out} can be calculated by inserting Eqs. 2.24 and 2.22 in Eq. 2.14

$$R_{\text{out,SC}} = \frac{\sum \left(\left(\dfrac{a_{\text{out}}^{(i)} \cdot q_{\text{out}}}{R_i C_i (1 - e^{-\tau_i/(R_i C_i)})} \right)^2 \dfrac{R_i^2 C_i}{2} (1 - e^{-2\tau_i/(R_i C_i)}) \right)}{f_{\text{sw}} \cdot \left(\sum \left(a_{\text{out}}^{(i)} \cdot q_{\text{out}} \right) \right)^2}. \tag{2.25}$$

The output resistance is a function of the frequency f_{sw} and the converter parameters R_i, C_i, and q_{out}. While Eq. 2.25 is generally applicable for all possible operation points, it can be simplified for the slow-switching limit (SSL) and fast-switching limit (FSL) as introduced in [72–75]. For $\tau_i \gg R_i C_i$, the SSL approximation can be expressed with

$$R_{\text{out,SSL}} = \frac{\sum \left(\dfrac{\left(a_{\text{out}}^{(i)} \right)^2}{C_i} \right)}{2 f_{\text{sw}} \left(\sum a_{\text{out}}^{(i)} \right)^2}. \tag{2.26}$$

In the slow-switching limit, the capacitive charge transfer losses dominate. At high switching frequencies, i.e., $\tau_i \ll R_i C_i$, the parasitic resistance dominates, and the equivalent output resistance can be approximated by

$$R_{\text{out,FSL}} = \frac{\sum \left(\frac{\left(a_{\text{out}}^{(i)} \right)^2 R_i}{\tau_i} \right)}{f_{\text{sw}} \left(\sum a_{\text{out}}^{(i)} \right)^2}. \tag{2.27}$$

Equations 2.26 and 2.27 are similar to the ones derived in [72]. R_i represents the summarized resistance value, while C_i is the effective capacitance value in each phase i.

Equivalent Output Resistance for Dynamic Off-Time Modulation

When the converter operates at the damped resonance frequency in each phase $\tau_i = T_{\text{res,d},i}/2$, the calculation of the current $I(t)$ (Eq. 2.17) and thus the equivalent output resistance (Eq. 2.14) can be significantly simplified. This operation mode is also called dynamic off-time modulation (DOTM) and is proposed in [39, 43, 44, 51, 52]. In [80] DOTM is implemented with an additional blocking diode, which prevents negative currents. Typical current waveforms are shown in Fig. 2.14. Solving Eq. 2.17 for the current and inserting in Eq. 2.14 lead to the following equation for the equivalent output resistance $R_{\text{out,DOTM}}$ [78–80].

$$R_{\text{out,DOTM}} = \frac{\sum \left(\frac{\left(a_{\text{out}}^{(i)} \right)^2}{C_i} \tanh \left(\frac{R_i T_{\text{res,d},i}}{8 L_i} \right) \right)}{2 f_{\text{sw}} \left(\sum a_{\text{out}}^{(i)} \right)^2} \tag{2.28}$$

Fig. 2.14 Current waveforms in the dynamic off-time operation mode

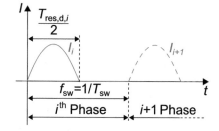

Fig. 2.15 Equivalent output
resistance R_{out} vs. switching
frequency f_{sw} for different
operation modes

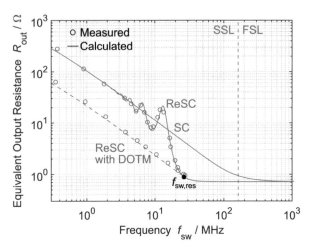

Equation 2.28 is valid for switching frequencies $f_{sw} \leq 1/T_{res,d,i}$. At higher switching frequencies, the parasitic resistance dominates, and the equivalent output resistance approaches the FSL resistance value.

Comparison of Different Options

Figure 2.15 shows a comparison of the R_{out} models for a SC converter (Eq. 2.25) and for a ReSC converter stage in full resonant behavior (Eq. 2.14) as well as with dynamic off-time operation (DOTM) (Eq. 2.28). The same switch resistance R_i and capacitor value C_i are used for comparison. The ReSC converter uses a small additional inductor L_i between the flying and the output capacitor as shown in Fig. 2.8a for resonant operation. The equivalent output resistance of the SC converter decreases linearly with increasing switching frequency since the reduced voltage ripple on the capacitors leads to lower charge transfer losses. At high switching frequencies, the equivalent output resistance is limited by the parasitic resistance in the power path, which is called the FSL limit $R_{out,FSL}$. The ReSC reaches this limit at much lower switching frequencies since the capacitive charge transfer losses are eliminated at the resonance frequency as explained in Sect. 2.2.4. The factor between output resistance of the SC and ReSC converter at the resonance frequency $f_{sw,res}$ is roughly the quality factor Q_i of the resonance circuit $(4/(\pi Q_i))$ formed by the flying capacitors and the inductance, damped by the switch and inductor resistance. More details can be found in [78, 79]. In DOTM operation mode, the output resistance decreases with increasing switching frequency similar to the SC curve but with significantly lower resistance value. The calculated values from the R_{out} model show a good agreement with the measurement results of the full resonant and DOTM operation. The measurement results originate from the implemented multi-ratio resonant SC converter described in Chap. 3.

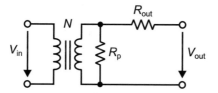

Fig. 2.16 Generalized equivalent output resistance model for SC converter and ReSC converter with additional parallel resistance R_p which models the extrinsic loss mechanisms

2.3.3 Extrinsic Loss Mechanisms

The extrinsic losses are mainly caused by the control of the switches and parasitic effects. They can be modeled by a parallel resistance R_p in the equivalent output resistance model shown in Fig. 2.16.

Switch Control Losses

The dominant part of the frequency-dependent losses are in most cases the gate charge losses that originate from the charging and discharging of the gate capacitances of the transistors. For the high-side switches, the gate charge is delivered by a charge pump or a bootstrap circuit with its corresponding efficiency (see Sect. 4.4). Detailed calculations of the switching losses are introduced in Sect. 4.1.2 where different implementation options for the power switches are modeled. For the proposed ReSC converter, a stack of two low-voltage 1.5 V transistors is chosen [81] for the implementation of the power switches since it has sufficient voltage blocking capability as described in Sect. 4.5.1. Therefore, the switching losses can be calculated with Eqs. 4.4 and 4.5. The total losses can be calculated with

$$P_{\text{sw,tot}} = \sum_{i=1}^{n} P_{\text{sw},i} + P_{\text{sw,CP},i} \tag{2.29}$$

where n is the number of switches in the power stage. $P_{\text{sw,CP},i}$ are the dynamic switching losses of the switch i in the power stage which are delivered by the charge pump, while $P_{\text{sw},i}$ are the switching losses delivered by the power stage itself (see Sect. 4.1.2). For the floating high-side switches, additional level shifters are used as described in Sect. 4.3. These are supplied by the charge pump with its efficiency η_{CP}. The power consumption of the level shifters $P_{\text{LS,tot}}$ can be extracted from simulation at an operation frequency f_x. Since the losses scale linearly with the switching frequency f_{sw}, the power consumption can be approximated with

$$P_{\text{LS,tot}} = \sum_{i=1}^{n} \left(\frac{P_{\text{LS},f_x}}{f_x} \cdot f_{\text{sw}} \right) \cdot \frac{1}{\eta_{\text{CP}}}. \tag{2.30}$$

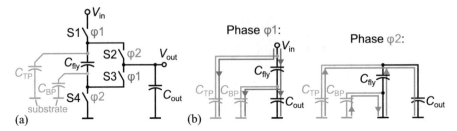

Fig. 2.17 (**a**) Parasitic capacitances at the flying capacitor; (**b**) charge flow in each phase with parasitic capacitances

Capacitive Bottom-Plate Losses

With the use of integrated capacitors, parasitic capacitances to the substrate are unavoidable. At the flying capacitors, which are responsible for the charge transport in the converter, charge sharing loss also occurs at these parasitic capacitances. The parasitic capacitances at the top plate C_{TP} and the bottom plate C_{BP} of a flying capacitor C_{fly} in a simple 1/2 SC cell are shown in Fig. 2.17a. During phase $\varphi1$, both capacitors C_{TP} and C_{BP} are charged, while in phase $\varphi2$ both capacitors are discharged (Fig. 2.17b). While the charge of the top-plate capacitor C_{TP} contributes to the output current, the charge of the bottom-plate capacitor C_{BP} is short-circuited to GND. Therefore, the parasitic bottom-plate capacitor C_{BP} is the main loss contributor. The quality of an integrated capacitor is expressed with the quality factor α and is defined as

$$\alpha = \frac{C_{BP}}{C_{fly}}. \tag{2.31}$$

The losses introduced by the bottom-plate capacitor C_{BP} can be calculated from

$$P_{CBP} = \sum_{i=1}^{n} \Delta V_{C_{BP},i}^2 \cdot C_{BP,i} \cdot f_{sw}. \tag{2.32}$$

$\Delta V_{C_{BP},i}$ is the voltage swing at the corresponding parasitic capacitor $C_{BP,i}$. In Sect. 5.1 different capacitor options in a standard CMOS process are analyzed, and the implementation of loss optimized capacitors is described.

References

1. Huang, J., et al.: WearDrive: fast and energy-efficient storage for wearables. In: 2015 USENIX Annual Technical Conference (USENIX ATC 15), pp. 613–625. USENIX Association, Santa Clara (2015). ISBN: 978-1-931971-225. https://www.usenix.org/conference/atc15/technical-session/presentation/huang-jian

2. Liu, R., Lin, F.X.: Understanding the characteristics of android wear OS. In: Proceedings of the 14th Annual International Conference on Mobile Systems, Applications, and Services, MobiSys'16, pp. 151–164. Association for Computing Machinery, Singapore (2016). ISBN: 9781450342698. https://doi.org/10.1145/2906388.2906398

3. Liu, X., Qian, F.: Measuring and optimizing android smartwatch energy consumption: poster. In: Proceedings of the 22nd Annual International Conference on Mobile Computing and Networking, MobiCom'16, pp. 421–423. Association for Computing Machinery, New York City (2016). ISBN: 9781450342261. https://doi.org/10.1145/2973750.2985259

4. Liu, X. et al.: Characterizing smartwatch usage in the wild. In: Proceedings of the 15th Annual International Conference on Mobile Systems, Applications, and Services, MobiSys'17, pp. 385–398. Association for Computing Machinery, Niagara Falls (2017a). ISBN: 9781450349284. https://doi.org/10.1145/3081333.3081351

5. Visuri, A., et al.: Quantifying sources and types of smartwatch usage sessions. In: Proceedings of the 2017 CHI Conference on Human Factors in Computing Systems, CHI'17, pp. 3569–3581. Association for Computing Machinery, Denver (2017). ISBN: 9781450346559. https://doi.org/10.1145/3025453.3025817

6. Linden, D., Reddy, T.B.: Handbook of Batteries, 3rd edn. McGraw-Hill (2002). ISBN: 0-07-135978-8

7. Ramadass, Y.K. Energy processing circuits for low-power applications. PhD thesis. Massachusetts Institute of Technology, Department of Electrical Engineering and Computer Science (2009)

8. Lee, S., et al.: A 0.518mm2 quasi-current-mode hysteretic buck DCDC converter with $3\mu s$ load transient response in $0.35\mu m$ BCDMOS. In: 2015 IEEE International Solid-State Circuits Conference – (ISSCC) Digest of Technical Papers, pp. 1–3 (2015). https://doi.org/10.1109/ISSCC.2015.7063002

9. Lee, S., et al.: Robust and efficient synchronous buck converter with near-optimal dead-time control. In: 2011 IEEE International Solid-State Circuits Conference, pp. 392–394 (2011). https://doi.org/10.1109/ISSCC.2011.5746366

10. Ahmad, H.H., Bakkaloglu, B.: A 300mA 14mV-ripple digitally controlled buck converter using frequency domain delta-sigma ADC and hybrid PWM generator. In: 2010 IEEE International Solid-State Circuits Conference – (ISSCC), pp. 202–203 (2010). https://doi.org/10.1109/ISSCC.2010.5433985

11. Paidimarri, A., Chandrakasan, A.P.: A buck converter with 240pW quiescent power, 92% peak efficiency and a 2×10^6 dynamic range. In: 2017 IEEE International Solid-State Circuits Conference (ISSCC), pp. 192–193 (2017). https://doi.org/10.1109/ISSCC.2017.7870326

12. Krishnamurthy, H.K., et al.: A digitally controlled fully integrated voltage regulator with on-die solenoid inductor with planar magnetic core in 14-nm tri-gate CMOS. IEEE J. Solid-State Circuits 53(1), 8–19 (2018a). ISSN: 0018-9200. https://doi.org/10.1109/JSSC.2017.2759117

13. Kudva, S.S., Harjani, R.: Fully-integrated on-chip DCDC converter with a $450\times$ output range. IEEE J. Solid-State Circuits 46(8), 1940–1951 (2011). ISSN: 0018-9200. https://doi.org/10.1109/JSSC.2011.2157253

14. Krishnamurthy, H.K., et al.: A 500MHz, 68% efficient, fully on-die digitally controlled buck voltage regulator on 22nm tri-gate CMOS. In: 2014 Symposium on VLSI Circuits Digest of Technical Papers, pp. 1–2 (2014). https://doi.org/10.1109/VLSIC.2014.6858438

15. Krishnamurthy, H.K., et al.: A digitally controlled fully integrated voltage regulator with 3-D-TSV-based on-die solenoid inductor with a planar magnetic core for 3-D-stacked die Applications in 14-nm tri-gate CMOS. IEEE J. Solid-State Circuits 53(4), 1038–1048 (2018b). ISSN: 1558-173X. https://doi.org/10.1109/JSSC.2017.2773637

16. Wens, M., Steyaert, M.S.J.: A fully integrated CMOS 800-mW four-phase semiconstant ON/OFF-time step-down converter. IEEE Trans. Power Electron. 26(2), 326–333 (2011). ISSN: 0885-8993. https://doi.org/10.1109/TPEL.2010.2057445

17. Wens, M., Steyaert, M.: A fully-integrated $0.18\mu m$ CMOS DCDC step-down converter, using a bondwire spiral inductor. In: 2008 IEEE Custom Integrated Circuits Conference, pp. 17–20 (2008). https://doi.org/10.1109/CICC.2008.4672009

18. Lee, M., Choi, Y., Kim, J.: A 0.76W/mm2 on-chip fully-integrated buck converter with negatively-coupled, stacked-LC filter in 65nm CMOS. In: 2014 IEEE Energy Conversion Congress and Exposition (ECCE), pp. 2208–2212 (2014). https://doi.org/10.1109/ECCE.2014. 6953697
19. Kyaw, P.A., Stein, A.L.F., Sullivan, C.R.: Fundamental examination of multiple potential passive component technologies for future power electronics. IEEE Trans. Power Electron. **33**(12), 10708–10722 (2018). ISSN: 1941-0107. https://doi.org/10.1109/TPEL.2017.2776609
20. Stauth, J.T.: Pathways to mm-scale DCDC converters: trends, opportunities, and limitations. In: 2018 IEEE Custom Integrated Circuits Conference (CICC), pp. 1–8 (2018). https://doi.org/ 10.1109/CICC.2018.8357017
21. Lutz, D., Renz, P., Wicht, B.: A 10mW fully integrated 2-to-13V-input buck-boost SC converter with 81.5% peak efficiency. In: 2016 IEEE International Solid-State Circuits Conference (ISSCC), pp. 224–225 (2016a). https://doi.org/10.1109/ISSCC.2016.7417988
22. Lutz, D., Renz, P., Wicht, B.: A 120/230Vrms-to-3.3V micro power supply with a fully integrated 17V SC DCDC converter. In: ESSCIRC Conference 2016: 42nd European Solid-State Circuits Conference, pp. 449–452 (2016b). https://doi.org/10.1109/ESSCIRC.2016.7598338
23. Lutz, D., Renz, P., Wicht, B.: An integrated 3-mW 120/230-V AC mains micropower supply. IEEE J. Emerg. Sel. Top. Power Electron. **6**(2), 581–591 (2018). ISSN: 2168-6777. https://doi. org/10.1109/JESTPE.2018.2798504
24. Sarafianos, A., Steyaert, M.: The folding Dickson converter: a step towards fully integrated wide input range capacitive DCDC converters. In: ESSCIRC 2014 – 40th European Solid State Circuits Conference (ESSCIRC), pp. 267–270 (2014). https://doi.org/10.1109/ESSCIRC.2014. 6942073
25. Sarafianos, A. et al.: A folding Dickson-based fully integrated wide input range capacitive DCDC converter achieving Vout/2-resolution and 71% average efficiency. In: 2015 IEEE Asian Solid-State Circuits Conference (A-SSCC), pp. 1–4 (2015). https://doi.org/10.1109/ASSCC. 2015.7387488
26. Bang, S., Blaauw, D., Sylvester, D.: A successive-approximation switched-capacitor DC–DC converter with resolution of $V_{IN}/2^N$ for a wide range of input and output voltages. IEEE J. Solid-State Circuits **51**(2), 543–556 (2016). https://doi.org/10.1109/JSSC.2015.2501985
27. Nguyen, B., et al.: High-efficiency fully integrated switched-capacitor voltage regulator for battery-connected applications in low-breakdown process technologies. IEEE Trans. Power Electron. **33**(8), 6858–6868 (2018). https://doi.org/10.1109/TPEL.2017.2757950.3
28. Salem, L.G., Mercier, P.P.: An 85%-efficiency fully integrated 15-ratio recursive switched-capacitor DCDC converter with 0.1-to-2.2V output voltage range. In: 2014 IEEE International Solid-State Circuits Conference Digest of Technical Papers (ISSCC), pp. 88–89 (2014). https:// doi.org/10.1109/ISSCC.2014.6757350
29. Butzen, N., Steyaert, M.S.J.: Scalable parasitic charge redistribution: design of high-efficiency fully integrated switched-capacitor DC–DC converters. IEEE J. Solid-State Circuits **51**(12), 2843–2853 (2016). ISSN: 1558-173X. https://doi.org/10.1109/JSSC.2016.2608349
30. Butzen, N., Steyaert, M.: Design of single-topology continuously scalable-conversion-ratio switched-capacitor DC–DC converters. IEEE J. Solid-State Circuits **54**(4), 1039–1047 (2019). ISSN: 1558-173X. https://doi.org/10.1109/JSSC.2018.2884351
31. Le, H., Sanders, S.R., Alon, E.: Design techniques for fully integrated switched-capacitor DCDC converters. IEEE J. Solid-State Circuits **46**(9), 2120–2131 (2011). ISSN: 1558-173X. https://doi.org/10.1109/JSSC.2011.2159054
32. Bang, S., et al.: A fully integrated successive-approximation switched-capacitor DCDC converter with 31mV output voltage resolution. In: 2013 IEEE International Solid-State Circuits Conference Digest of Technical Papers, pp. 370–371 (2013). https://doi.org/10.1109/ ISSCC.2013.6487774
33. Liu, W., et al.: A 94.2%-peak-efficiency 1.53A direct-battery-hook-up hybrid Dickson switched-capacitor DCDC converter with wide continuous conversion ratio in 65nm CMOS. In: 2017 IEEE International Solid-State Circuits Conference (ISSCC), pp. 182–183 (2017b). https://doi.org/10.1109/ISSCC.2017.7870321

34. Kim, W., Brooks, D., Wei, G.: A fully-integrated 3-level DCDC converter for nanosecond-scale DVFS. IEEE J. Solid-State Circuits **47**(1), 206–219 (2012). ISSN: 0018-9200. https://doi.org/10.1109/JSSC.2011.2169309

35. Hardy, C., Le, H.: A 10.9W 93.4%-efficient (27W 97%-efficient) flying-inductor hybrid DCDC converter suitable for 1-cell (2-cell) battery charging applications. In: 2019 IEEE International Solid-State Circuits Conference – (ISSCC), pp. 150–152 (2019). https://doi.org/10.1109/ISSCC.2019.8662432

36. Schaef, C., Stauth, J.T.: A 3-phase resonant switched capacitor converter delivering 7.7W at 85% efficiency using 1.1nH PCB trace inductors. IEEE J. Solid-State Circuits **50**(12), 2861–2869 (2015). ISSN: 1558-173X. https://doi.org/10.1109/JSSC.2015.2462351

37. Schaef, C., Kesarwani, K., Stauth, J.T.: A variable-conversion-ratio 3-phase resonant switched capacitor converter with 85% efficiency at $0.91W/mm^2$ using 1.1nH PCB-trace inductors. In: 2016 IEEE International Solid-State Circuits Conference – (ISSCC) Digest of Technical Papers, pp. 1–3 (2015). https://doi.org/10.1109/ISSCC.2015.7063075

38. Schaef, C., Din, E., Stauth, J.T.: A digitally controlled 94.8%-peak-efficiency hybrid switched-capacitor converter for bidirectional balancing and impedance-based diagnostics of lithium-ion battery arrays. In: 2017 IEEE International Solid-State Circuits Conference (ISSCC), pp. 180–181 (2017). https://doi.org/10.1109/ISSCC.2017.7870320

39. Renz, P., Lueders, M., Wicht, B.: A 47 MHz Hybrid Resonant SC Converter with Digital Switch Conductance Regulation and Multi-Mode Control for Li-Ion Battery Applications, 2020 IEEE Applied Power Electronics Conference and Exposition (APEC), New Orleans, LA, USA, 2020, pp. 15–18, https://doi.org/10.1109/APEC39645.2020.9124238

40. Renz, P., et al.: A fully integrated 85%-peak-efficiency hybrid multi ratio resonant DCDC converter with 3.0-to-4.5V input and $500\mu A$-to-120mA load range. In: 2019 IEEE International Solid-State Circuits Conference – (ISSCC), pp. 156–158 (2019a). https://doi.org/10.1109/ISSCC.2019.8662491

41. Renz, P., et al.: A 3-ratio 85% efficient resonant SC converter with on-chip coil for Li-ion battery operation. IEEE Solid-State Circuits Lett. **2**(11), 236–239 (2019b). ISSN: 2573-9603. https://doi.org/10.1109/LSSC.2019.2927131

42. Kesarwani, K., Stauth, J.T.: The direct-conversion resonant switched capacitor architecture with merged multiphase interleaving: cost and performance comparison. In: 2015 IEEE Applied Power Electronics Conference and Exposition (APEC), pp. 952–959 (2015a). https://doi.org/10.1109/APEC.2015.7104464

43. Kesarwani, K., Stauth, J.T.: Resonant and multi-mode operation of flying capacitor multi-level DCDC converters. In: 2015 IEEE 16th Workshop on Control and Modeling for Power Electronics (COMPEL), pp. 1–8 (2015b). https://doi.org/10.1109/COMPEL.2015.7236511

44. Kesarwani, K., Sangwan, R., Stauth, J.T.: A 2-phase resonant switched-capacitor converter delivering 4.3W at 0.6W/mm2 with 85% efficiency. In: 2014 IEEE International Solid-State Circuits Conference Digest of Technical Papers (ISSCC), pp. 86–87 (2014). https://doi.org/10.1109/ISSCC.2014.6757349

45. Kudva, S., Chaubey, S., Harjani, R.: High power-density, hybrid inductive/capacitive converter with area reuse for multi-domain DVS. In: Proceedings of the IEEE 2014 Custom Integrated Circuits Conference, pp. 1–4 (2014). https://doi.org/10.1109/CICC.2014.6946053

46. Baek, J., et al.: Switched inductor capacitor buck converter with >85% power efficiency in 100uA-to-300mA loads using a bang-bang zero-current detector. In: 2018 IEEE Custom Integrated Circuits Conference (CICC), pp. 1–4 (2018). https://doi.org/10.1109/CICC.2018.8357024

47. Pilawa-Podgurski, R.C.N., Giuliano, D.M., Perreault, D.J.: Merged two-stage power converter architecture with softcharging switched-capacitor energy transfer. In: 2008 IEEE Power Electronics Specialists Conference, pp. 4008–4015 (2008). https://doi.org/10.1109/PESC.2008.4592581

48. Pilawa-Podgurski, R.C.N., Perreault, D.J.: Merged two-stage power converter with soft charging switched-capacitor stage in 180nm CMOS. IEEE J. Solid-State Circuits **47**(7), 1557–1567 (2012). ISSN: 1558-173X. https://doi.org/10.1109/JSSC.2012.2191325

49. Sun, J., et al.: High power density, high efficiency system two-stage power architecture for laptop computers. In: 2006 37th IEEE Power Electronics Specialists Conference, pp. 1–7 (2006). https://doi.org/10.1109/pesc.2006.1711768

50. Sun, J., Xu, M., Lee, F.C.: Transient analysis of the novel voltage divider. In: APEC 07 – Twenty-Second Annual IEEE Applied Power Electronics Conference and Exposition, pp. 550–556 (2007). https://doi.org/10.1109/APEX.2007.357568

51. Cheng, K.W.E.: New generation of switched capacitor converters. In: PESC 98 Record. 29th Annual IEEE Power Electronics Specialists Conference (Cat. No.98CH36196), vol. 2, pp. 1529–1535 (1998). https://doi.org/10.1109/PESC.1998.703377

52. Lin, Y.-C., Liaw D.-C.: Parametric study of a resonant switched capacitor DCDC converter. In: Proceedings of IEEE Region 10 International Conference on Electrical and Electronic Technology. TENCON 2001 (Cat. No.01CH37239), vol. 2, pp. 710–716 (2001). https://doi.org/10.1109/TENCON.2001.949684

53. Qiu, D., Zhang, B., Zheng, C.: Duty ratio control of resonant switched capacitor DCDC converter. In: 2005 International Conference on Electrical Machines and Systems, vol. 2, pp. 1138–1141 (2005). https://doi.org/10.1109/ICEMS.2005.202724

54. Liu, X., et al.: Analysis and design considerations of integrated 3-level buck converters. IEEE Trans. Circuits Syst.: Regular Papers **63**(5), 671–682 (2016). ISSN: 1558-0806. https://doi.org/10.1109/TCSI.2016.2556098

55. Godycki, W., Sun, B., Apsel, A.: Part-time resonant switching for light load efficiency improvement of a 3-level fully integrated buck converter. In: ESSCIRC 2014 – 40th European Solid State Circuits Conference (ESSCIRC), pp. 163–166 (2014). https://doi.org/10.1109/ESSCIRC.2014.6942047

56. Villar, G., Alarcon, E.: Monolithic integration of a 3-level DCM-operated low-floating-capacitor buck converter for DCDC step-down conversion in standard CMOS. In: 2008 IEEE Power Electronics Specialists Conference, pp. 4229–4235 (2008). https://doi.org/10.1109/PESC.2008.4592620

57. Kumar, P., et al.: A 0.4V-1V 0.2A/mm2 70% efficient 500MHz fully integrated digitally controlled 3-level buck voltage regulator with on-die high density MIM capacitor in 22nm tri-gate CMOS. In: 2015 IEEE Custom Integrated Circuits Conference (CICC), pp. 1–4 (2015). https://doi.org/10.1109/CICC.2015.7338479

58. Kiani, M.H., Stauth, J.T.: Optimization and comparison of hybrid-resonant switched capacitor DCDC converter topologies. In: 2017 IEEE 18th Workshop on Control and Modeling for Power Electronics (COMPEL), pp. 1–8 (2017). https://doi.org/10.1109/COMPEL.2017.8013321

59. Schaef, C., Stauth, J.T.: A 12-volt-input hybrid switched capacitor voltage regulator based on a modified series-parallel topology. In: 2017 IEEE Applied Power Electronics Conference and Exposition (APEC), pp. 2453–2458 (2017). https://doi.org/10.1109/APEC.2017.7931043

60. Abdulslam, A., Mercier, P.P.: A symmetric modified multilevel ladder PMIC for battery-connected applications. IEEE J. Solid-State Circuits, 1–14 (2019a). ISSN: 1558-173X. https://doi.org/10.1109/JSSC.2019.2957658

61. Li, Y., et al.: AC-coupled stacked dual-active-bridge DCDC converter for integrated lithium-ion battery power delivery. IEEE J. Solid-State Circuits **54**(3), 733–744 (2019). ISSN: 1558-173X. https://doi.org/10.1109/JSSC.2018.2883746

62. Macy, B.B., Lei, Y., Pilawa-Podgurski, R.C.N.: A 1.2 MHz, 25 V to 100 V GaN-based resonant Dickson switched-capacitor converter with 1011 W/in3 (61.7 kW/L) power density. In: 2015 IEEE Applied Power Electronics Conference and Exposition (APEC), pp. 1472–1478 (2015). https://doi.org/10.1109/APEC.2015.7104542

63. Ye, Z., Lei, Y., Pilawa-Podgurski, R.C.N.: A resonant switched capacitor based 4-to-1 bus converter achieving 2180 W/in3 power density and 98.9% peak efficiency. In: 2018 IEEE Applied Power Electronics Conference and Exposition (APEC), pp. 121–126 (2018). https://doi.org/10.1109/APEC.2018.8340997

64. Meynard, T.A., Foch, H.: Multi-level conversion: high voltage choppers and voltage-source inverters. In: PESC'92Record. 23rd Annual IEEE Power Electronics Specialists Conference, vol. 1, pp. 397–403 (1992). https://doi.org/10.1109/PESC.1992.254717

65. Amin, S.S., Mercier, P.P.: A fully integrated Li-ion-compatible hybrid four-level DC–DC converter in 28-nm FDSOI. IEEE J. Solid-State Circuits **54**(3), 720–732 (2019). ISSN: 0018-9200. https://doi.org/10.1109/JSSC.2018.2880183

66. Abdulslam, A., Ismail, Y.: 5-level buck converter with reduced inductor size suitable for on-chip integration. In: 5th International Conference on Energy Aware Computing Systems Applications, pp. 1–4 (2015). https://doi.org/10.1109/ICEAC.2015.7352205

67. Lei, Y., et al.: A 2 kW, single-phase, 7-level, GaN inverter with an active energy buffer achieving 216 W/in3 power density and 97.6% peak efficiency. In: 2016 IEEE Applied Power Electronics Conference and Exposition (APEC), pp. 1512–1519 (2016). https://doi.org/10.1109/APEC.2016.7468068

68. Rentmeister, J.S., et al.: A flying capacitor multilevel converter with sampled valley-current detection for multi-mode operation and capacitor voltage balancing. In: 2016 IEEE Energy Conversion Congress and Exposition (ECCE), pp. 1–8 (2016). https://doi.org/10.1109/ECCE.2016.7854680

69. Rentmeister, J.S., Stauth, J.T.: A 48V:2V flying capacitor multilevel converter using current-limit control for flying capacitor balance. In: 2017 IEEE Applied Power Electronics Conference and Exposition (APEC), pp. 367–372 (2017). https://doi.org/10.1109/APEC.2017.7930719

70. Ripamonti, G., et al.: An integrated regulated resonant switched-capacitor DCDC converter for PoL applications. In: 2019 IEEE Applied Power Electronics Conference and Exposition (APEC), pp. 207–211 (2019). https://doi.org/10.1109/APEC.2019.8722136

71. Abdulslam, A., Mercier, P.P.: A continuous-input-current passive-stacked third-order buck converter achieving 0.7W/mm2 power density and 94% peak efficiency. In: 2019 IEEE International Solid-State Circuits Conference – (ISSCC), pp. 148–150 (2019b). https://doi.org/10.1109/ISSCC.2019.8662384

72. Seeman, M.D.: A design methodology for switched-capacitor DCDC converters. PhD thesis. EECS Department, University of California, Berkeley (2009). https://www2.eecs.berkeley.edu/Pubs/TechRpts/2009/EECS-2009-78.html

73. Makowski, M.S., Maksimovic, D.: Performance limits of switched-capacitor DCDC converters. In: Proceedings of PESC'95 – Power Electronics Specialist Conference, vol. 2, pp. 1215–1221 (1995). https://doi.org/10.1109/PESC.1995.474969

74. Arntzen, B., Maksimovic, D.: Switched-capacitor DC/DC converters with resonant gate drive. In: IEEE Trans. Power Electron. **13**(5), 892–902 (1998). ISSN: 1941-0107. https://doi.org/10.1109/63.712304

75. Kimball, J.W., Krein, P.T.: Analysis and design of switched capacitor converters. In: Twentieth Annual IEEE Applied Power Electronics Conference and Exposition, 2005. APEC 2005, vol. 3, pp. 1473–1477 (2005). https://doi.org/10.1109/APEC.2005.1453227

76. Salem, L., Ismail, Y.: Slow-switching-limit loss removal in SC DCDC converters using adiabatic charging. In: 2011 International Conference on Energy Aware Computing, pp. 1–4 (2011). https://doi.org/10.1109/ICEAC.2011.6136700

77. Salem, L., Ismail, Y.: Switched-capacitor DCDC converters with output inductive filter. In: 2012 IEEE International Symposium on Circuits and Systems (ISCAS), pp. 444–447 (2012). https://doi.org/10.1109/ISCAS.2012.6272059

78. Pasternak, S., Schaef, C., Stauth, J.: Equivalent resistance approach to optimization, analysis and comparison of hybrid/resonant switched-capacitor converters. In: 2016 IEEE 17th Workshop on Control and Modeling for Power Electronics (COMPEL), pp. 1–8 (2016). https://doi.org/10.1109/COMPEL.2016.7556737

79. Pasternak, S.R., et al.: Modeling and performance limits of switched-capacitor DCDC converters capable of resonant operation with a single inductor. IEEE J. Emerg. Sel. Top. Power Electron. **5**(4), 1746–1760 (2017). ISSN: 2168-6785. https://doi.org/10.1109/JESTPE.2017.2730823

80. Evzelman, M., Ben-Yaakov, S.: Average-current-based conduction losses model of switched capacitor converters. IEEE Trans. Power Electron. **28**(7), 3341–3352 (2013). ISSN: 1941-0107. https://doi.org/10.1109/TPEL.2012.2226060
81. Renz, P., Kaufmann, M., Lueders, M., Wicht, B.: Switch stacking in power management ICs. IEEE J. Emerg. Sel. Top. Power Electron. (JESTPE) (2020, In press). https://doi.org/10.1109/JESTPE.2020.3012813

Chapter 3
Multi-Ratio Resonant Switched-Capacitor Converters

Resonant SC operation offers the highest potential for full integration with high conversion efficiency as derived in Sect. 2.2.5. This is mainly because of the lower switching frequencies and good component utilization. In this chapter, *multi-ratio* resonant operation is applied in order to cover a wide input voltage range. To support a wide output power range, different control mechanisms are introduced and compared with each other.

3.1 Multi-Ratio Resonant Conversion

In order to cover the wide input voltage range of Li-ion batteries of 3.0–4.5 V, the basic 1/2 ratio ReSC topology shown in Fig. 2.8a is extended as shown in Fig. 3.1. The standard 1/2 cell is placed two times (S1–S4 and S6–S9) with one additional switch between the two flying capacitors C_{fly1} and C_{fly2}. In addition to $N = 1/2$, this configuration enables two more voltage conversion ratios, $N = 2/3$ and $N = 1/3$ (Fig. 3.1). The target output voltage V_{out} is set between 1.5 and 1.8 V and can be scaled to 1.2 V and lower. Depending on the input and output voltage, the three ratios offer a lossless coarse control of the output voltage V_{out} by an outer control loop described in Sect. 6.1.3. In addition, the fine regulation adapts the equivalent output resistance R_{out}. Different options are discussed in Sect. 3.2. The power stage consists of nine switches S1–S9, which operate at different voltage levels during each switching phase. For highly efficient operation, stacked low-voltage switches are used as described in Sect. 4.1. This requires several level shifters (see Sect. 4.3) and a gate drive supply generation for each switch (see Sect. 4.4). Low-power design of the additional circuits is crucial to keep the efficiency impact as small as possible.

Multi-phase time-interleaving is a widely used technique in SC converters [1–3], which reduces the amount of input current ripple and output voltage ripple. Thus, the output buffer capacitor can be reduced or even removed [3]. In conventional SC

© The Editor(s) (if applicable) and The Author(s), under exclusive license to
Springer Nature Switzerland AG 2021
P. Renz, B. Wicht, *Integrated Hybrid Resonant DCDC Converters*,
https://doi.org/10.1007/978-3-030-63944-0_3

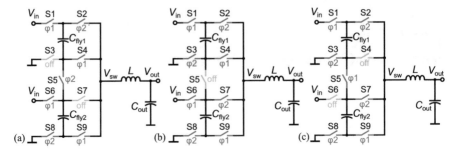

Fig. 3.1 Proposed multi-ratio resonant SC power stage: (**a**) ratio $N = 2/3$; (**b**) ratio $N = 1/2$; (**c**) ratio $N = 1/3$

converters, a segmentation of the flying capacitors for time-interleaved operation is possible without degrading the quality factor of each. In resonant SC converters, however, the extent of interleaving is limited since each phase requires an inductor. Segmentation of the inductor leads to a significant higher resonance frequency in each phase, and thus higher switching losses, degrading the overall efficiency. Additional losses caused by control overhead circuits like level shifters and gate drive supply generation may further degrade the efficiency. This is the reason why no interleaving approach is applied in this work.

Different switch configurations for each conversion ratio of the proposed ReSC converter are shown in Fig. 3.2. In the 1/2 ratio (Fig. 3.1b), the two flying capacitors C_{fly1} and C_{fly2} are always in parallel in phase $\varphi 1$ and $\varphi 2$. The resonance frequency $f_{sw,res,1/2} = 1/T_{res,1/2}$ is defined by

$$f_{sw,res,1/2} = \left(2\pi\sqrt{L(C_{fly1} + C_{fly2})}\right)^{-1} \tag{3.1}$$

with a duty cycle of $D_{1/2} = T_{\varphi 1}/T_{res,1/2} = 1/2$.

In the 1/3 and 2/3 ratios, the flying capacitors are connected in series during one of the phases leading to a higher resonance frequency and different duty cycles as indicated by the waveforms in Fig. 3.2a and c.

$$f_{sw,res,2/3} = f_{sw,res,1/3} = \pi\sqrt{L}\left(\sqrt{C_{fly1} + C_{fly2}} + \sqrt{\frac{C_{fly1}C_{fly2}}{C_{fly1} + C_{fly2}}}\right)^{-1}. \tag{3.2}$$

With $C_{fly1} = C_{fly2}$, this results in a factor 4/3 higher resonance frequency compared to ratio 1/2

$$f_{sw,res,2/3} = f_{sw,res,1/3} = \frac{4}{3} \cdot f_{sw,res,1/2}. \tag{3.3}$$

In addition, different duty cycles of $D_{2/3} = T_{\varphi 1}/T_{res,2/3} = 2/3$ for the ratio 2/3 and $D_{1/3} = T_{\varphi 1}/T_{res,1/3} = 1/3$ for the ratio 1/3 are required for resonant

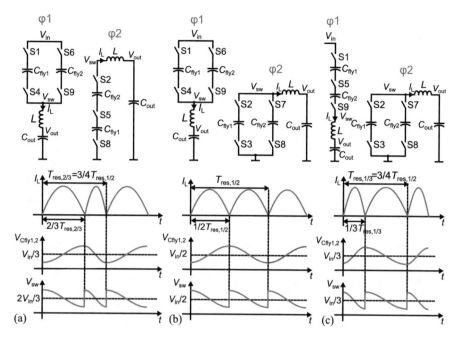

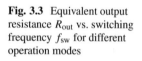

Fig. 3.2 Proposed multi-ratio resonant operation: (a) ratio $N = 2/3$; (b) ratio $N = 1/2$; (c) ratio $N = 1/3$

Fig. 3.3 Equivalent output resistance R_{out} vs. switching frequency f_{sw} for different operation modes

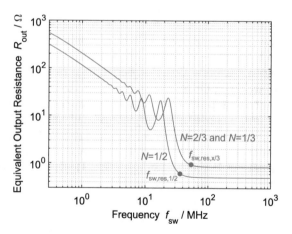

operation. The resonance frequencies and duty cycles are automatically adjusted by a switchable current source in the relaxation oscillator designed as part of this work (see Sect. 6.1.5).

The equivalent output resistance R_{out} for the three different conversion ratios is shown in Fig. 3.3 according to Eq. 2.14 (Sect. 2.3.2). The flying capacitors are set to $C_{fly1} = C_{fly2} = 1\,\text{nF}$, the inductor is set to $L = 10\,\text{nH}$, and the switches are designed with $R_{sw} = 500\,m\Omega$ (see Sect. 3.3). According to Eqs. 3.1 and 3.2, this leads to resonance frequencies of $f_{sw,res,1/2} = 35.58\,MHz$ and $f_{sw,res,2/3} = f_{sw,res,1/3} = 47.45\,MHz$, which matches the R_{out} calculations in Fig. 3.3. Due to the complementary switching phases in the 1/3 and 2/3 conversion ratios, the output resistances are identical for both ratios. In ratio 1/2, a FSL resistance of $R_{out,FSL} = 500\,m\Omega$ is obtained. A slightly higher FSL resistance of $R_{out,FSL} = 830\,m\Omega$ can be observed in ratio 1/3 and 2/3 due to series connection of three switches either in phase $\varphi 1$ (ratio 1/3) or in $\varphi 2$ (ratio 2/3). The calculation of the equivalent component values R_i, C_i for the corresponding switching phase i is shown in Appendix.

Even more conversion ratios could be beneficial for an efficient coverage of the wide input voltage range (see Fig. 2.6). As shown in Fig. 3.4a, a conversion ratio 4/7 can be inserted between 2/3 and 1/2. Based on simulations for switch conductance regulation (SwCR), introduced in Sect. 3.2.2, this increases the converter efficiency between 3.3 and 3.7 V.

Conventionally, SC and ReSC converters operate in two switching phases. In that case, a higher number of ratios N leads to a significantly increasing complexity and chip area, since topology reconfiguration requires more switches and flying capacitors [4–6]. In ReSC converters this would lead to a higher resistance in the power path, resulting in a lower-quality factor of the resonance circuit. Dividing the switching period into three or more phases enables new possibilities [7–10]. In contrast to two-phase converters, this technique provides additional ratios with fewer flying capacitors and switches, making it attractive for high-quality resonant operation.

The proposed 4/7 conversion ratio can be implemented by the use of only three flying capacitors instead of a minimum of four, which would be required in a two-phase topology. The existing power stage of Fig. 3.1 has to be extended by an additional flying capacitor C_{fly3} as shown in Fig. 3.4b. In this design, the converter operates at a switching frequency of 45.5 MHz and $C_{fly3} = 250\,pF$. Further information about the 4/7 conversion ratio can be found in Appendix.

The simulated efficiency versus input voltage in Fig. 3.4a shows that the 4/7 ratio leads to an efficiency improvement of up to 7%. Compared to the 2/3 and 1/2 ratio, a lower peak efficiency of 81% is observed, due to a slightly higher equivalent output resistance in three-phase operation. However, a more flat efficiency curve across the Li-ion input voltage range is achieved. In the framework of this research, the ratio 4/7 was not implemented as priority was given to the multi-ratio converter shown in Fig. 3.1, various supporting circuits, and different control methods. As a future work, the conversion ratio 4/7 is worth to be implemented and experimentally verified. Because of the low component count, this concept is particularly suitable for resonant operation, and further investigations on even more phases and even higher input voltages are worthwhile.

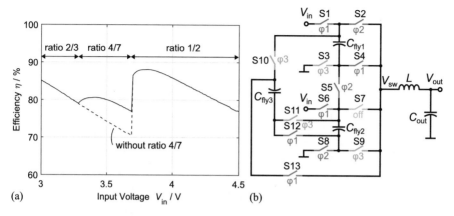

Fig. 3.4 Implementation of additional conversion ratio 4/7: (**a**) simulated converter efficiency η versus input voltage V_{in} with $V_{out} = 1.8\,\text{V}$, $C_{fly1} = C_{fly2} = 1\,\text{nF}$, $C_{fly3} = 250\,\text{pF}$, $L = 10\,\text{nH}$; (**b**) power stage implementation

3.2 Control Mechanisms

The conversion ratios of the ReSC converter offer a lossless coarse control of the output voltage. For the fine regulation, the equivalent output resistance R_{out} of the converter needs to be modified. Comparable to a linear regulator, the resistance R_{out} needs to be adapted depending on the input voltage V_{in}, the conversion ratio N of the SC cell, the desired output voltage V_{out}, and the load current I_{out}, as expressed by Eq. 3.4 (see also Fig. 2.13c)

$$R_{out} = (N \cdot V_{in} - V_{out})\,\frac{1}{I_{out}}. \tag{3.4}$$

There are several ways to modify the equivalent output resistance R_{out}, but all of them introduce additional losses in the power path. For the overall goal of high efficiency, the adaption of R_{out} should be performed in a way that other extrinsic losses (see Sect. 2.3.3) are reduced. These losses mainly consists of the gate charge losses $P_{sw,tot}$ as defined in Eq. 2.29 and bottom-plate losses of the integrated flying capacitors C_{fly1} and C_{fly2} according to Eq. 2.32. Both loss components are directly proportional to the switching frequency f_{sw}.

Different approaches for modulating the equivalent output resistance R_{out} are shown in Fig. 3.5. The most obvious method to adapt R_{out} is to change the switching frequency f_{sw} of the converter around the resonance frequency $f_{sw,res}$ (operating range (1) in Fig. 3.5; see also Eq. 2.14). But, due to the steep slope of R_{out}, the converter still operates at high frequencies leading to excessive switching losses, which degrade the efficiency. The occurrence of a negative inductor current is a further disadvantage, which causes additional losses and does not allow zero current switching. These disadvantages can be overcome by the dynamic off-time modula-

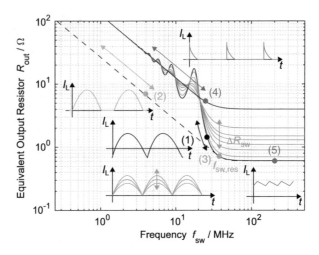

Fig. 3.5 Equivalent output resistance R_{out} versus switching frequency f_{sw} for different control approaches for ReSC converters: (1) frequency modulation around resonance point; (2) dynamic off-time modulation (DOTM); (3) ReSC operation with SwCR at resonance frequency $f_{sw,res}$; (4) SC operation with frequency modulation; (5) inductive PWM regulation

tion approach (DOTM) (operating range (2) in Fig. 3.5, described in Sect. 3.2.1). Switch conductance regulation (SwCR) at constant resonance frequency $f_{sw,res}$ (operating range (3) in Fig. 3.5) offers zero current switching and low output voltage ripple, even with small passives. It also achieves a flat converter efficiency over a wide load range. SwCR is further introduced in Sect. 3.2.2. SC converters achieve good performance at low-power levels [4, 5, 11–14]. Therefore, SC operation with frequency modulation at low output currents (operating range (4) in Fig. 3.5) can also be beneficial for resonant SC converters and is investigated in Sect. 3.2.2. All of the introduced control methods adjust the equivalent output resistance R_{out} and introduce additional losses in the power path. Nevertheless, this control method, in combination with the resonant multi-ratio approach, is preferred over inductive PWM control (operating range (5) in Fig. 3.5), since it leads to significant lower switching frequencies and therefore higher overall efficiency (see Sect. 2.2.5), especially for the proposed fully integrated converter.

3.2.1 Dynamic Off-Time Modulation (DOTM)

The dynamic off-time modulation approach (DOTM) is introduced in [15–19] (operation range (2) in Fig. 3.5). The idea is to reduce the effective switching frequency f_{sw} of the converter while maintaining the benefits of the resonant charging and zero current switching. To achieve this, the conduction time of the switches is fixed at the resonance frequency $f_{sw,res}$ (see Eqs. 3.1 and 3.2), but a dead

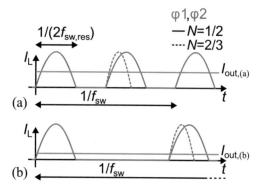

Fig. 3.6 Inductor current in DOTM control for (**a**) an arbitrary load current $I_{\text{out},(a)}$ and (**b**) a different load current $I_{\text{out},(b)} < I_{\text{out},(a)}$

time is added, where all switches are off. This dead time is modulated according to the present load current and voltage conditions. The resulting inductor current I_L waveforms are presented in Fig. 3.6.

The output current I_{out} can be calculated by

$$I_{\text{out}} = 2 f_{\text{sw}} \int_0^{\frac{1}{2 f_{\text{sw,res}}}} I_L(t) \, dt = \frac{f_{\text{sw}}}{f_{\text{sw,res}}} \cdot \frac{2}{\pi} \cdot \hat{I}_L, \qquad (3.5)$$

while the RMS current $I_{\text{out,RMS}}$ is defined as

$$I_{\text{out,RMS}} = \sqrt{2 f_{\text{sw}} \int_0^{\frac{1}{2 f_{\text{sw,res}}}} I_L^2(t) \, dt} = \sqrt{\frac{f_{\text{sw}}}{f_{\text{sw,res}}}} \cdot \frac{1}{\sqrt{2}} \cdot \hat{I}_L. \qquad (3.6)$$

Equation 3.5 shows that the output current I_{out} scales linearly with the ratio between the resonant pulse and off-time ($f_{\text{sw,res}}/f_{\text{sw}}$), while the RMS current $I_{\text{out,RMS}}$ scales only with the square root of this ratio. Since the RMS current corresponds to the losses in the converter, this results in an increasing equivalent output resistance R_{out} with decreasing output current or switching frequency f_{sw}. As shown in Fig. 3.5, this leads to an R_{out} curve in parallel to the curve of the SC converter (see Sect. 2.3.2). The gap between the two curves is caused by the lack of charge sharing losses due to resonant charging in DOTM mode.

From the efficiency point of view, DOTM is an appropriate method for the regulation of the output voltage since it directly scales down the frequency-dependent losses. However, the discontinuous current flow leads to a large output voltage ripple ΔV_{out} with small integrated buffer capacitors, especially when a higher R_{out} is needed for regulation purposes. For example, at an input voltage of $V_{\text{in}} = 4.5 \, \text{V}$, an unacceptable output voltage ripple of $\Delta V_{\text{out}} > 500 \, \text{mV}$ is obtained for a small integrated output capacitor of $C_{\text{out}} = 10 \, \text{nF}$ ($C_{\text{fly1}} = C_{\text{fly2}} = 1 \, \text{nF}$,

$L = 10\,\text{nH}$, $V_{\text{out}} = 1.8\,\text{V}$, $I_{\text{out}} = 50\,\text{mA}$). To reduce the ripple of the output voltage to an acceptable level, an output capacitance C_{out} in the range of $100\,\text{nF}$ is required, which contradicts the idea of a fully integrated voltage converter. The DOTM control method is also implemented in the proposed work with an additional external output capacitor (see Sects. 3.4 and 6.1.2). The converter in [17] is implemented with integrated buffer capacitors; however, it operates only near the ideal conversion ratio.

3.2.2 Switch Conductance Regulation (SwCR) and SC Mode

In order to achieve a low output voltage ripple with integrated passives, operation at high resonance frequency $f_{\text{sw,res}}$ and modulation of the switch conductance ($1/R_{\text{sw}}$) for adjusting the equivalent output resistance R_{out} are proposed (operating range (3) in Fig. 3.5). As a main benefit, the proposed converter offers zero current switching and efficient resonant charging. While DOTM is adapting the off-time between similar, high-energy pulses, SwCR operates at constant resonance frequency and changes the transferred energy of each resonant pulse. Therefore, a small output capacitor of $C_{\text{out}} = 10\,\text{nF}$ is sufficient to achieve a satisfying low output voltage ripple ΔV_{out}. This raises the SwCR to the best suited regulation method for the proposed fully integrated ReSC voltage converter.

There are several ways to control the on-resistance of a MOS transistor used as a switch in the converter. The drain current equation results in an on-resistance $R_{\text{DS,on}}$ for an NMOStransistor in the linear region of

$$R_{\text{DS,on}} = \left(\mu_{\text{n}} C_{\text{ox}} \cdot \frac{W}{L} (V_{\text{GS}} - V_{\text{th}}) \right)^{-1}. \tag{3.7}$$

Modulation of the switch resistance $R_{\text{DS,on}}$ while scaling down the gate charge losses for highly efficient converter operation can be done by varying the gate-source voltage V_{GS} or the width W of a transistor. Figure 3.1 shows that most of the switches in the proposed power stage are high-side switches, where at least one terminal changes its potential at each switching phase. Controlling the $R_{\text{DS,on}}$ by changing the gate-source voltage V_{GS} would require a precise control for the gate voltage of each transistor at flying potential with very fast transitions. Since this approach is very complex, the converter in [20] uses an additional switch at the output node whose gate-source voltage is then regulated depending on the output voltage V_{out}, similar to a series-connected LDO. As the major disadvantage of this approach, the gate charge losses do not scale proportionally since all switches in the power stage are operated at a high frequency with their original size, leading to high gate charge losses.

Scaling the gate charge losses of each power switch toward lower output power can be achieved by binary segmentation and weighting of the power switches as indicated in Fig. 3.7. The width W of the power switches is adjusted during

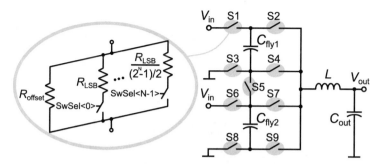

Fig. 3.7 Basic implementation of the segmented power switches for SwCR control

operation by a digital control loop with reasonable design overhead. The control signals can be generated globally in the low-side domain. Afterward, they can be transferred to the high-side domain via level shifters (see Sect. 4.3).

At lower output currents I_{out}, there is no advantage in further reducing the transistor width W. On the one hand, this would require a very large dynamic range for the width W scaling; on the other hand, there is no benefit of maintaining the high resonance frequency. At low output currents I_{out}, a high equivalent output resistance R_{out} and therefore also a high switch resistance R_{sw} are required (see Eq. 3.4). In consequence, as illustrated in Fig. 3.5, the R_{out} curve approaches the SC curve (operating range (4) in Fig. 3.5) since the resonances are more and more damped. Therefore, at low output power, the regulation is done by modulating the switching frequency f_{sw}, which is referred to as SC mode. There, the inductor is ineffective. Lower f_{sw} also reduces capacitive bottom-plate losses (see Sect. 2.3.3) and thus maintains high efficiency at low power. In SC mode, all switch segments are off, except for one remaining offset segment. Both operation modes are marked in Fig. 3.8, where the equivalent output resistance R_{out} is plotted over the switching frequency f_{sw} and the switch resistance R_{sw} of the power switches. The mode transition from SwCR to SC mode is performed automatically by the control as described in Sect. 6.1.

The maximum required switch resistance $R_{sw,max} = R_{offset}$ (Fig. 3.7) can be calculated with the damping factor ζ of the resonant circuit in the corresponding phase $i = 1, 2$.

$$\zeta = \frac{1}{2Q} = \frac{R_{sw}}{2} \cdot \sqrt{\frac{C_i}{L}} \tag{3.8}$$

For a damping factor of $\zeta > 1$, the system is overdamped, and the converter can be considered as a SC converter where the inductor L is ineffective. In the final implementation, a switch resistance of $R_{offset} = 6\,\Omega$ is chosen, which results in a damping factor of $\zeta = 1.3$.

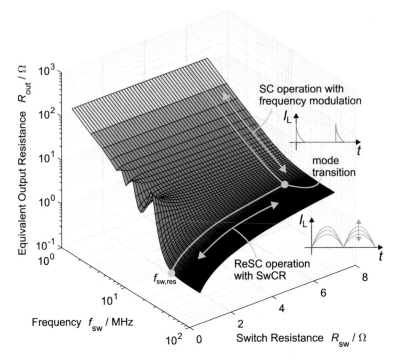

Fig. 3.8 Equivalent output resistance R_{out} versus switching frequency f_{sw} and switch resistance R_{sw}

To achieve a high-resolution switch resistance with minimum design overhead, the power switches are segmented into binary weighted parts. This leads to a quantization error of the output voltage, which corresponds to an additional V_{out} ripple. Figure 3.9 shows the output voltage V_{out} versus the output current I_{out} for different width resolutions is shown in Fig. 3.9, based on a MATLAB® model of the control. The plot is created for the worst-case situation with an input voltage of $V_{in} = 4.5$ V, which corresponds to a higher resistance value R_{sw}. Not all of the possible resistance values can be used for regulation purposes, which leads to a lower effective resolution and, therefore, to a higher quantization error. To maintain a V_{out} control deviation of less than 10 mV, a resolution of 8 bits is required. Therefore, the transistors are divided into nine segments, one offset transistor and eight binary weighted transistors, which are controlled with the 8-bit SwSel<0:7> signal. Implementation details can be found in Chap. 4, Sect. 4.2.

For SwSel<0:7>= 0, all segments are turned off, except the offset resistance, which leads to a switch resistance of $R_{sw,max} = 6\,\Omega$. For SwSel<0:7>= 255, all segments are turned on, leading to a minimum resistance value of $R_{sw,min} = 500$ mΩ. More information on $R_{sw,min}$ can be found in Sect. 3.3. Due to the parallel arrangement of the resistors, it is more practical to use the switch conductance G_{sw},

Fig. 3.9 Output voltage V_{out} over output current I_{out} for different width resolutions

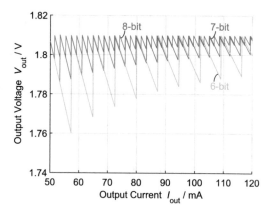

Fig. 3.10 Switch resistance R_{sw} of segmented power switch over digital control code Sw<0:7>

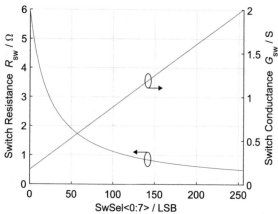

which leads to the linear relationship versus the digital control value SwSel<0:7>, plotted in Fig. 3.10.

$$G_{sw} = G_{LSB} \cdot SwSel + G_{offset}, \tag{3.9}$$

Accordingly, the switch resistance $R_{sw} = 1/G_{sw}$ graph forms a rectangular hyperbola. This fits very well with Eq. 3.4 ($R_{out} = f(I_{out})$), which also shows a hyperbolic behavior. Hence, there is also a linear relationship between the digital control value SwSel<0:7> and the output current I_{out}.

Chapter 4 shows the implementation details of the power stage, while the integration of passive components is discussed in Sect. 5. Details of the modeling and implementation of different control loops for DOTM, SwCR, and SC are given in Chap. 6.

3.3 Efficiency Modeling and Optimization

A general figure of merit (FOM) model is introduced in [21] where different hybrid converter topologies are compared in terms of their respective performance limits. The series-parallel structure, which is used in this work, has the best capacitor utilization and thus leads to lower resonance frequencies. This relationship was confirmed for SC converters in [22]. However, according to the results in [21], there is little difference among various topologies for moderate conversion ratios up to $N = 8$.

In this section, an efficiency model is derived for the determination of an optimal design point. It contains all relevant design parameters: flying capacitors $C_{fly1,2}$, inductor L, resonance frequency $f_{sw,res}$, and output power P_{out}. The corresponding optimum design point is extracted for full resonant operation at high output power in the conversion ratio $N = 1/2$. There, the highest efficiency can be expected since it has the lowest equivalent output resistance (see Fig. 3.4). For further investigation and analysis, other converter operation and control modes (SwCR, SC, and DOTM) are also included in the efficiency model.

The efficiency η of a DCDC converter can be calculated in general form by

$$\eta = \frac{P_{out}}{P_{in}} = \frac{P_{out}}{P_{out} + P_{loss}}. \tag{3.10}$$

The equivalent output resistance R_{out} model of Sect. 2.3.2 forms the basis for the converter efficiency model. Additional to the intrinsic R_{out} losses ($P_{Rout} = R_{out} \cdot I_{out}^2$), the frequency-dependent extrinsic loss mechanisms have to be included, i.e., switch control losses $P_{sw,tot}$ and $P_{LS,tot}$ (Sect. 2.3.3) as well as capacitive bottom-plate losses P_{CBP} (Sect. 2.3.3), which yields

$$P_{loss} = P_{Rout} + P_{sw,tot} + P_{LS,tot} + P_{CBP}. \tag{3.11}$$

Any static DC power losses are neglected in the first step. They can be added later and do not affect the optimization process. According to the charge flow analysis (see Apendix 3.4), the flying capacitors are sized equally ($C_{fly1} = C_{fly2}$) since the charge flow elements of both are identical in all ratios. The same applies for the power switches (S1–S9).

In order to determine the optimal design point of the converter, the efficiency η is plotted versus inductor L and total flying capacitor value C_{fly} as shown in Fig. 3.11a. C_{fly} is $C_{fly} = C_{fly1} + C_{fly2}$ in ratio $N = 1/2$. The converter operates in full resonant operation at an input voltage of $V_{in} = 3.8$ V and an output current of $I_{out} = 120$ mA. The resonance frequency $f_{sw,res}$ is set according to the C_{fly} and L values (Eq. 3.1). The power switches are designed at their sweet spot (see analysis in Sect. 4.1.2) with an on-resistance of $R_{sw} = 500$ mΩ. Metal-insulator-metal MIM capacitors (see Sect. 5.1) are assumed in the efficiency model for the calculation of the losses and of the area consumption. For the implementation of the inductor, integrated

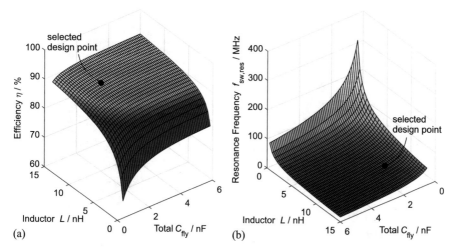

Fig. 3.11 Determination of optimum design point for resonant operation (at $V_{in} = 3.8$ V $I_{out} = 120$ mA, $R_{sw} = 500$ mΩ): (**a**) efficiency η versus inductor L and total flying capacitor value $C_{fly} = C_{fly1} + C_{fly2}$; (**b**) resonance frequency $f_{sw,res}$ versus inductor L and total flying capacitor value C_{fly}

planar air-core inductors are used, as described in Sect. 5.2.2. Figure 3.11a reveals that larger values of both the flying capacitors and the inductor lead to a higher overall efficiency. This is mainly due to the fact that the resonance frequency $f_{sw,res}$ decreases significantly as shown in Fig. 3.11b, leading to lower frequency-dependent switching losses. Increasing inductance and capacitance values also lead to a higher-quality factor Q of the resonance circuit as shown in Fig. 3.12a, but at the cost of a higher layout area if C_{fly} and L are fully integrated (Fig. 3.12b).

In Figs. 3.11 and 3.12, the selected design points are marked, corresponding to values of $C_{fly} = 2$ nF and $L = 10$ nH. It shows the best trade-off among high peak efficiency $\eta = 89\%$, moderate resonance frequency $f_{sw,res} = 35.58$ MHz, quality factor $Q = 4.5$, and small area consumption $A_{Cfly} + A_L = 4$ mm^2 of the flying capacitors and the inductor.

Figure 3.13 shows the loss distribution of the converter over the total flying capacitor C_{fly} and over the inductor L value. With increasing C_{fly}, the switching losses $P_{sw,tot} + P_{LS,tot}$ decrease, but this is counteracted by the increasing capacitive bottom-plate losses P_{CBP}. The losses due to the equivalent output resistance P_{Rout} are constant as they only depend on the switch resistance R_{sw} and the equivalent series resistance R_{ESR} of the inductor. The selected design point at $C_{fly} = 2$ nF shows a good trade-off between the total power losses P_{loss} and the overall area consumption. An increasing inductor value L leads to lower switching losses $P_{sw,tot} + P_{LS,tot}$ and capacitive bottom-plate losses P_{CBP}, as shown in Fig. 3.13b. An inductance value of $L = 10$ nH was chosen for the selected operating point. There is no benefit in using higher inductor values since it leads to a higher equivalent

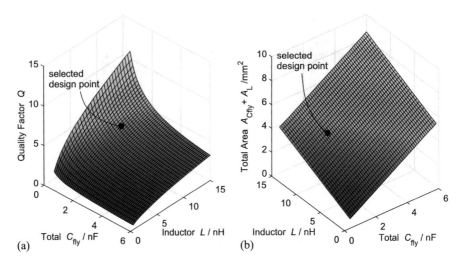

Fig. 3.12 Determination of optimum design point for resonant operation (at $V_{\text{in}} = 3.8\,\text{V}$, $I_{\text{out}} = 120\,\text{mA}$, $R_{\text{sw}} = 500\,\text{m}\Omega$): (**a**) quality factor Q of the C_{fly}-L circuit versus inductor L and total flying capacitor value $C_{\text{fly}} = C_{\text{fly1}} + C_{\text{fly2}}$; (**b**) area consumption of the integrated flying capacitors $A_{C\text{fly}}$ and inductor A_L versus inductor L and total flying capacitor value C_{fly}

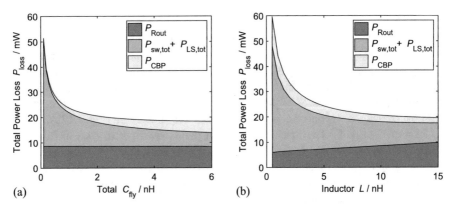

Fig. 3.13 Loss distribution for resonant operation (at $V_{\text{in}} = 3.8\,\text{V}$, $I_{\text{out}} = 120\,\text{mA}$, $R_{\text{sw}} = 500\,\text{m}\Omega$): (**a**) total power loss P_{loss} versus flying capacitor value C_{fly} for $L = 10\,\text{nH}$; (**b**) total power loss P_{loss} versus inductor value L for $C_{\text{fly}} = 2\,\text{nF}$

series resistance R_{ESR} of the inductor, which, in turn, leads to higher P_{Rout} losses. Furthermore, it would also lead to a significant higher area consumption.

For further investigations, the efficiency model is extended by different control modes: switch conductance regulation (SwCR), SC mode operation, and dynamic off-time modulation (DOTM). These are modeled by the different expressions for the equivalent output resistance R_{out}, introduced in Sect. 2.3.2, which are adjusted according to the control variables, i.e., switch resistance R_{sw} for SwCR control

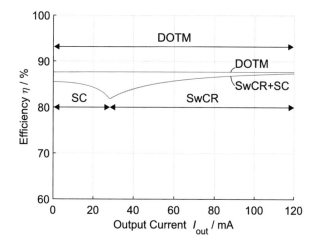

Fig. 3.14 Calculated efficiency η versus output current I_{out} for different control approaches with $V_{in} = 3.9\,\text{V}$, $V_{out} = 1.8\,\text{V}$, $C_{fly1} = C_{fly2} = 1\,\text{nF}$, $L = 10\,\text{nH}$

and switching frequency f_{sw} for SC and DOTM control. Figure 3.14 shows the efficiency η of the different control methods over the output current I_{out}. At high output currents, the efficiency is the same for both regulation methods, DOTM and SwCR. This is expected since the inductor current I_L waveforms are identical for both regulation methods at high currents. Moving to medium load currents, the conversion with DOTM regulation maintains a nearly constant efficiency, while the efficiency of the SwCR control method reduces slightly since the capacitive bottom-plate losses become a dominant part of the losses at the constant high switching frequency . At low output currents, SC operation with modulation of the switching frequency f_{sw} reduces the bottom-plate losses and thus maintains high efficiency at low power (see Sect. 3.2.2). Although, the operation in SwCR and SC mode leads to a slightly lower efficiency at medium output power, it enables the use of an on-chip output capacitor $C_{out} = 10\,\text{nF}$, leading to a small output voltage ripple $<50\,\text{mV}$. Simulations have shown that the output voltage ripple in DOTM mode can reach large values of $\Delta V_{out} > 500\,\text{mV}$ for the worst-case condition $V_{in} = 4.5\,\text{V}$. For this reason an external output capacitor of $C_{out} = 100\,\text{nF}$ has to be used in DOTM mode.

The accuracy of the efficiency model for the implemented resonant SC converter is shown in Fig. 3.15. Figure 3.15a shows the results of the fully integrated option with SwCR and SC control. The efficiency model tracks the measurement results with an error of less than 5% in both operating modes. In Fig. 3.15b, the efficiency model for DOTM mode and the measurement results match with 4% accuracy. A slightly higher peak efficiency compared to SwCR operation can be observed since it was measured with an external 10 nH inductor, which has a lower equivalent series resistor R_{ESR}.

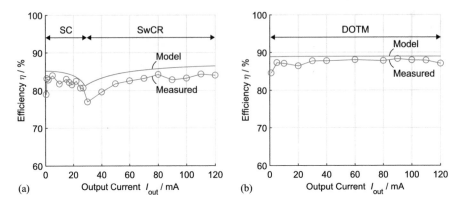

Fig. 3.15 Comparison between efficiency model and measurement results. Efficiency η versus output current I_{out} for different control approaches: (**a**) SwCR and SC mode with integrated spiral inductor $L = 10\,\text{nH}$ and output capacitor $C_{out} = 10\,\text{nF}$; (**b**) DOTM mode with off-chip inductor placed in the package $L = 10\,\text{nH}$ (Coilcraft 0603HP-10NXGLU) and external output capacitor $C_{out} = 100\,\text{nF}$

3.4   Experimental Results

The resonant SC converter has been manufactured in a $130\,\text{nm}$ BCD process. Figure 3.16a shows the die micrograph of the implemented converter. The chip has an active area of $7\,\text{mm}^2$ and is assembled in a $5 \times 5\,\text{mm}$ QFN package. The flying capacitors C_{fly1} and C_{fly2} are implemented using MIM capacitors with a capacitance of $1\,\text{nF}$ each (see Sect. 5.1.1). A version with MOS capacitors was also implemented (not shown in Fig. 3.16). A comparison of the measured converter efficiency with MIM and MOS capacitor options (MIM or MOS) is presented in Sect. 5.1.3. The output capacitor C_{out} is implemented as a MOS capacitor with $C_{out} = 10\,\text{nF}$. The input capacitor C_{in} consists of $10\,\text{nF}$ off-chip in parallel to a $180\,\text{pF}$ on-chip MIM capacitor.

Different options of passives are investigated depending on the regulation mode. The fully integrated option in Fig. 3.16a is implemented with an on-chip planar spiral inductor of $L = 10\,\text{nH}$. It has a DC resistance of $R_{ESR,DC} = 280\,\text{m}\Omega$ and a quality factor of $Q = 6$ at a frequency of $f = 35.5\,\text{MHz}$ (see Sect. 5.2.2). The highly integrated option is shown in Fig. 3.16b with a $10\,\text{nH}$ SMD inductor placed on top of the die. Figure 3.16a and b utilizes the SwCR and SC mode control. Figure 3.16c shows a discrete option for DOTM control where the inductor L and the output capacitor C_{out} are placed externally. The lower half of the die is not used, only the power stage with the flying capacitors and the control circuits. This reduces the required chip area by half to approximately $3.5\,\text{mm}^2$. The on-chip flying capacitors C_{fly1} and C_{fly2} are still used.

The measured efficiency η of the SwCR and SC control in Fig. 3.17 shows a flat run over the output current I_{out}, which is due to effective scaling of the switching losses by the proposed SwCR. High efficiency at very low power is maintained by

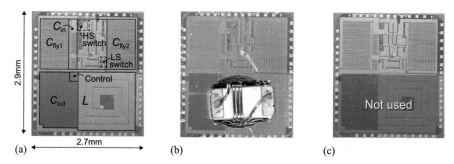

(a)   (b)   (c)

Fig. 3.16 Die micrographs of the test chip implemented in a 130 nm BCD process: (**a**) fully integrated option with integrated spiral inductor $L = 10$ nH; (**b**) highly integrated option with off-chip inductor placed in the package $L = 10$ nH (Coilcraft 0603HP-10NXGLU); (**c**) discrete option for DOTM control with external inductor $L = 10$ nH (Coilcraft 0603HP-10NXGLU) and output capacitor $C_{out} = 100$ nF

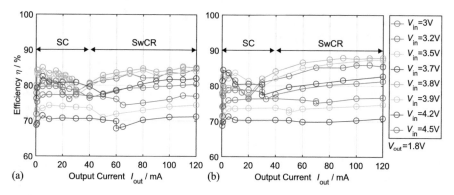

Fig. 3.17 Measured efficiency η versus output current I_{out} with SwCR and SC control: (**a**) fully integrated option with integrated spiral inductor $L = 10$ nH; (**b**) highly integrated option with off-chip inductor placed in the package $L = 10$ nH (Coilcraft 0603HP-10NXGLU)

dual-mode operation, which enters SC mode below 20–30 mA load. The design achieves a peak efficiency of 85% with the fully integrated planar inductor and 88.5% with a 10 nH in-package inductor at power densities of 0.033 W/mm^2 and 0.054 W/mm^2, respectively. The converter operates at resonance switching frequencies of up to 35.5 MHz (1/2 ratio) and 47.5 MHz ($x/3$ ratios).

The output voltage ripple ΔV_{out} for different conversion ratios is shown in Fig. 3.18a. As expected, a small output voltage ripple <35 mV for SwCR and <50 mV in low-power SC mode is achieved. Waveforms of the switching node V_{sw} and the clock signals $\varphi 1$ and $\varphi 2$, shown in Fig. 3.18b, confirm the resonant operation, in line with the curves in Fig. 3.2.

Figure 3.19 shows the measured efficiency η for DOTM control. In Fig. 3.19a, DOTM control with an external 10 nH inductor and 100 nF output capacitor indicates a peak efficiency of $\eta = 89\%$, similar to the SwCR option in Fig. 3.17b.

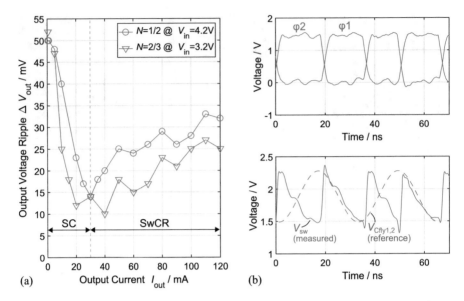

Fig. 3.18 Measured output voltage ripple and transient waveforms: (**a**) measured output voltage ripple ΔV_{out} versus output current I_{out} with SwCR and SC control for the fully integrated option; (**b**) measurement of transient waveforms of the switching node V_{sw} and clock signals $\varphi 1$ and $\varphi 2$

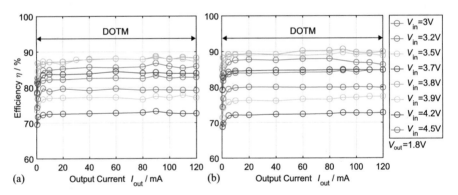

Fig. 3.19 Measured efficiency η versus output current I_{out} with DOTM control: (**a**) with off-chip inductor $L = 10$ nH (Coilcraft 0603HP-10NXGLU) and 100 nF off-chip output capacitor; (**b**) with off-chip inductor $L = 39$ nH (Coilcraft 0603HP-39NXJLU) and 100 nF off-chip output capacitor

In DOTM control, all frequency-dependent losses scale down effectively, which leads to a very flat efficiency curve over the output current I_{out}. The use of a higher inductance value of 39 nH results in a higher peak efficiency of $\eta = 91\%$ as indicated in Fig. 3.19b. This is mainly due to the lower resonance frequency $f_{sw,res} = 18$ MHz.

Tables 3.1 and 3.2 summarize the performance of the proposed converter options with SwCR and SC control in comparison to other designs of fully and highly

Table 3.1 Comparison of fully integrated option to state-of-the-art publications

	[28]	[27]	[25]	[26]	[11]	[12]	This work
				Fully integrated converters			
V_{in}/V	2.4	1.8	1.5	2.8–4.2	3.4–4.3	3.2–4.0	3.0–4.5
V_{out}/V	0.4–1.4	0.7	0.9–1.15	0.6–1.2	0.45–1.5	1	1.5–1.8
I_{out}/mA	300–850	16–200	30–330	16e-3 – 66	1e-3 – 300e-3	6–45	0.5–120
f_{sw}/MHz	50–200	N/A	100	200	2.7	10–100	35–50
C_{fly}	18 nF	1.5 nF	N/A	10 nF	2.24	720 pF	2 nF
C_{out}	10 nF	1.9 nF	5 nF	50 nF	n.r.	300 pF	10 nF
Resonant operation	no	no	no	no	no	no	yes
Technology	130 nm	65 nm	14 nm	28 nm	180 nm	130 nm	130 nm
Power density/mW/mm^2	200	67	1.15e3	26.7	0.27	120	33
L_{total}	4 nH	3 nH	1.5 nH	3 nH	N/A	N/A	9 nH
η_p/%	77	64	84	78	72	80	85
EEF[a] @ η_p/%	40.5	39.2	8.7	64	68	66.2	45.7
Load current dynamic range[b]	2.83	12.5	11	200	100	7.5	240

[a] EEF $= 1 - (\eta_{LDO}/\eta_{conv})$
[b] $I_{out,max} / I_{out,min}$ with $\eta > 0.8 \cdot \eta_p$ over V_{in} range

Table 3.2 Comparison of highly integrated option with state-of-the-art publications

	Highly integrated converters			
	[23]	[17]	[24]	**This work**
V_{in}/V	3.7–5	6/3.6	2.4–4.4[c]	3.0–4.5
V_{out}/V	1.2–2.5	3/1.8	1.0–2.2[c]	1.5–1.8
I_{out}/mA	500–1e3	50–1.2e3	n.r.	0.5–120
f_{sw}/MHz	20–50	20–40	47.5	35–50
C_{fly}	24 nF	18 nF	3.4 nF	2 nF
C_{out}	6 nF	11 nF	100 nF (ext)	10 nF
Resonant operation	yes	yes	yes	yes
Technology	180 nm	180 nm	180 nm	130 nm
Power density/mW/mm^2	0.91e3	0.6e3	97	54
L_{total}	3x5.5 nH (ext)	11 nH (ext)	7.7	10 nH (ext)
$\eta_p/\%$	85	85	85.5	88.5
EEF[a] @ $\eta_p/\%$	42.7	41.2	45.8	47.9
Load current dynamic range[b]	3	24	n.r.	240

[a]$EEF = 1 - (\eta_{LDO}/\eta_{conv})$
[b]$I_{out,max}/I_{out,min}$ with $\eta > 0.8 \cdot \eta_p$ over V_{in} range
[c]Full V_{in} and V_{out} range is not disclosed

integrated converters. All highly integrated converters use small inductors placed on top of the chip [17, 23] or an external output capacitor [24]. The converter of this work achieves the highest peak efficiency of 85% (fully integrated) and 88.5% (highly integrated). Compared to fully integrated inductively operated converters [25–28], significantly higher peak efficiency and input voltage ranges can be reached due to the multi-ratio resonant operation at lower switching frequencies. High efficiency enhancement factors (EEF), which are the improvement over an ideal linear regulator (see Eq. 2.1), of 45.7% (fully integrated) and 47.9% (highly integrated) are reached. Due to the multi-mode SC operation at low power, high efficiency is maintained over a wide output current range which leads to a large value of 240 for the dynamic load current range (see Eq. 2.4). The wide input voltage range is fully suitable for Li-ion operation.

In Fig. 3.20, several options of the converters proposed in this work are compared to state-of-the-art hybrid converters as presented in Sect. 2.2.5 (Fig. 2.10). The key metrics for wearable and IoT applications are reflected in Fig. 3.20: peak efficiency η_p, dynamic load current range $I_{out,max}/I_{out,min}$, and compatibility with Li-Ion input voltage range. The proposed fully integrated option is able to cover the full Li-ion input voltage range, which was not addressed by fully integrated state-of-the-art converters previously. It also outperforms other fully integrated inductive converters in terms of peak efficiency and dynamic load current range. Compared to highly integrated resonant SC or hybrid SC converters, a high peak efficiency in combination with a wide dynamic load current range and Li-ion compatibility is achieved. Therefore it is fully suitable for wearable and IoT applications.

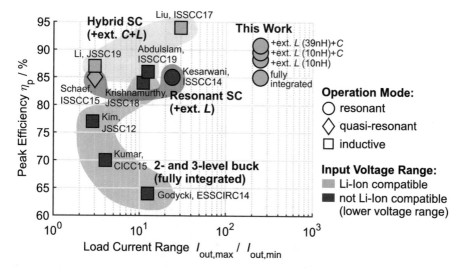

Fig. 3.20 Efficiency versus dynamic load current range of published highly integrated state-of-the-art hybrid DCDC converters [17, 23, 25, 27–31] in comparison with this work [19, 32–34]

Appendix

Charge Flow Analysis for Different Conversion Ratios

Figure 3.21a–c shows the different capacitor configurations for the separate phases $\varphi 1$ and $\varphi 2$ of the proposed multi-ratio ReSC converter depending on the conversion ratio N. For each phase i, a charge multiplier vector can be defined describing the topology based on the charge flow through the capacitors.

$$\boldsymbol{a}^{(i)} = \left[q_{\text{out}}^{(i)} \; q_{C_1}^{(i)} \; \cdots \; q_{C_n}^{(i)} \; q_{\text{in}}^{(i)}\right]^{\text{T}} \cdot \frac{1}{q_{\text{out}}} = \left[a_{\text{out}}^{(i)} \; a_{C_1}^{(i)} \; \cdots \; a_{C_n}^{(i)} \; a_{\text{in}}^{(i)}\right]^{\text{T}} \quad (3.12)$$

With Fig. 3.21a–c, the charge multiplier vectors of the different conversion ratios can be determined:
Ratio $N = 1/3$

$$\boldsymbol{a}_{1/3}^{(1)} = [q_x \; q_x \; q_x \; q_x]^{\text{T}} \cdot \frac{1}{3q_x} = \left[\frac{1}{3} \; \frac{1}{3} \; \frac{1}{3} \; \frac{1}{3}\right]^{\text{T}} \quad (3.13)$$

$$\boldsymbol{a}_{1/3}^{(2)} = [2q_x \; -q_x \; -q_x \; 0]^{\text{T}} \cdot \frac{1}{3q_x} = \left[\frac{2}{3} \; -\frac{1}{3} \; -\frac{1}{3} \; 0\right]^{\text{T}} \quad (3.14)$$

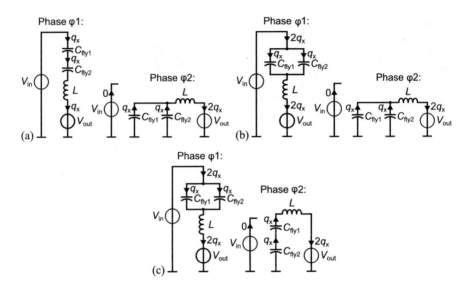

Fig. 3.21 Charge flow analysis for the resonant multi-ratio ReSC converter: (**a**) conversion ratio 1/3; (**b**) conversion ratio 1/2; (**c**) conversion ratio 2/3

Ratio N = 1/2

$$a_{1/2}^{(1)} = [2q_x \ q_x \ q_x \ 2q_x]^{\mathrm{T}} \cdot \frac{1}{4q_x} = \left[\frac{1}{2} \ \frac{1}{4} \ \frac{1}{4} \ \frac{1}{2}\right]^{\mathrm{T}} \tag{3.15}$$

$$a_{1/2}^{(2)} = [2q_x \ -q_x \ -q_x \ 0]^{\mathrm{T}} \cdot \frac{1}{4q_x} = \left[\frac{1}{2} \ -\frac{1}{4} \ -\frac{1}{4} \ 0\right]^{\mathrm{T}} \tag{3.16}$$

Ratio N = 2/3

$$a_{2/3}^{(1)} = [2q_x \ q_x \ q_x \ 2q_x]^{\mathrm{T}} \cdot \frac{1}{3q_x} = \left[\frac{2}{3} \ \frac{1}{3} \ \frac{1}{3} \ \frac{2}{3}\right]^{\mathrm{T}} \tag{3.17}$$

$$a_{2/3}^{(2)} = [2q_x \ -q_x \ -q_x \ 0]^{\mathrm{T}} \cdot \frac{1}{3q_x} = \left[\frac{1}{3} \ -\frac{1}{3} \ -\frac{1}{3} \ 0\right]^{\mathrm{T}} \tag{3.18}$$

For the calculation of the equivalent output resistance R_{out}, only the corresponding charge flow of the output has to be considered

$$\text{Ratio } N = 1/3: \quad q_{\mathrm{out},1/3}^{(1)} = \frac{1}{3} \quad q_{\mathrm{out},1/3}^{(1)} = \frac{2}{3}$$

$$\text{Ratio } N = 1/2: \quad q_{\mathrm{out},1/2}^{(1)} = \frac{1}{2} \quad q_{\mathrm{out},1/2}^{(2)} = \frac{1}{2}$$

$$\text{Ratio } N = 1/2: \quad q_{\mathrm{out},2/3}^{(1)} = \frac{2}{3} \quad q_{\mathrm{out},2/3}^{(2)} = \frac{1}{3}$$

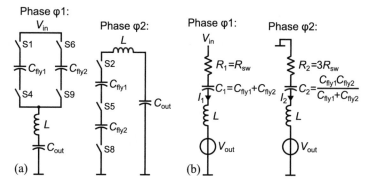

Fig. 3.22 (a) Schematic of the individual switching phases for 2/3 conversion ratio; (b) resulting equivalent circuits of the individual phases for calculation of the equivalent output resistance

Calculation of the Equivalent Output Resistance for Different Conversion Ratios

Determination of the Equivalent Circuits for the Switching Phases

To set up the differential equations for the current $I_i(t)$ according to the equivalent circuit for the corresponding phase as shown in Fig. 2.13b, it is necessary to determine the component values R_i, C_i for the corresponding switching phase i. This procedure is illustrated in the following for the conversion ratio 2/3. Figure 3.22a shows the resulting schematics for the phases $\varphi 1$ and $\varphi 2$ for the 2/3 conversion ratio. In phase $\varphi 1$, both flying capacitors C_{fly1} and C_{fly2} are connected in parallel, which leads to a resulting capacitor value of (with $C_{fly1} = C_{fly2} = C_{fly}$)

$$C_1 = C_{fly1} + C_{fly2} = 2 \cdot C_{fly}. \tag{3.19}$$

In phase $\varphi 2$ they are connected in series, resulting in

$$C_2 = \frac{C_{fly1} \cdot C_{fly2}}{C_{fly1} + C_{fly2}} = \frac{C_{fly}}{2}. \tag{3.20}$$

The power switches are connected in parallel in phase $\varphi 1$, which leads to an effective resistance R_1 of (assuming same on-resistance R_{sw} for all power switches)

$$R_1 = \frac{2R_{sw} \cdot 2R_{sw}}{2R_{sw} + 2R_{sw}} = R_{sw}. \tag{3.21}$$

In phase $\varphi 2$, the series connection of the power switches results in

$$R_2 = 3R_{sw} \tag{3.22}$$

Accordingly, this procedure can also be applied to the other conversion ratios.

Derivation of the Current Equations for the ReSC Converter

The equivalent circuit in Fig. 2.13b leads to a homogeneous second-order differential equation with constant coefficients.

$$0 = LC_i \cdot \frac{d^2 I_i}{dt^2} + R_i C_i \cdot \frac{dI_i}{dt} + I_i \qquad (3.23)$$

This can be solved easily to get the following generalized solution for the current $I(t)$

$$I(t) = \begin{cases} A_{11} \cdot e^{\lambda_{11}t} + A_{12} \cdot e^{\lambda_{12}t} & t \in [0, \tau_1] \\ A_{21} \cdot e^{\lambda_{21}(t-\tau_1)} + A_{22} \cdot e^{\lambda_{22}(t-\tau_1)} & t \in [\tau_1, \tau_1 + \tau_2] \end{cases} \qquad (3.24)$$

where the eigenvalues λ_{ik} are

$$\lambda_{ik} = \frac{-R_i}{2L} \pm \sqrt{\left(\frac{R_i}{2L}\right)^2 - \frac{1}{LC_i}}. \qquad (3.25)$$

For the determination of the constants A_{ik}, four additional boundary conditions are required. The first two can be determined with the charge flow analysis. The charge $q_{out}^{(i)}$ flowing into the output in each phase i must be equal to the integral of the current over the phase τ_i

$$q_{out}^{(i)} = a_{out}^{(i)} \cdot q_{out} = \int_0^{\tau_i} I_i(t) dt \qquad (3.26)$$

Phase $\varphi 1$:

$$a_{out}^{(1)} \cdot q_{out} = \frac{A_{11}}{\lambda_{11}} \cdot (e^{\lambda_{11}\tau_1} - 1) + \frac{A_{12}}{\lambda_{12}} \cdot (e^{\lambda_{12}\tau_1} - 1) \qquad (3.27)$$

Phase $\varphi 2$:

$$a_{out}^{(2)} \cdot q_{out} = \frac{A_{21}}{\lambda_{21}} \cdot (e^{\lambda_{21}\tau_1} - 1) + \frac{A_{22}}{\lambda_{22}} \cdot (e^{\lambda_{22}\tau_1} - 1) \qquad (3.28)$$

The other two boundary conditions can be determined by the transition during the phases. Due to the inductor, the current at the beginning of each switching phase must be the same as at the end of the previous one. For two switching phases ($i = 2$), this means

$$I_{1,2}(t = 0) = I_{2,1}(t = \tau_{2,1}) \qquad (3.29)$$

First transition

$$I_1(t = 0) = I_2(t = \tau_2) \tag{3.30}$$

$$A_{11} + A_{12} = A_{21} \cdot e^{\lambda_{21}\tau_2} + A_{22} \cdot e^{\lambda_{22}\tau_2} \tag{3.31}$$

Second transition

$$I_2(t = 0) = I_1(t = \tau_1) \tag{3.32}$$

$$A_{21} + A_{22} = A_{11} \cdot e^{\lambda_{11}\tau_1} + A_{12} \cdot e^{\lambda_{12}\tau_1} \tag{3.33}$$

Equations 3.27, 3.28, 3.31, and 3.33 lead to a 4×4 equation system for the determination of A_{11} to A_{22}. For the sake of clarity, the auxiliary variable W_{ik} is introduced

$$W_{ik} = \frac{e^{\lambda_{ik}\tau_i} - 1}{\lambda_{ik}}. \tag{3.34}$$

With that, the equation system can be written with

$$\begin{pmatrix} W_{11} & W_{12} & 0 & 0 \\ 0 & 0 & W_{21} & W_{22} \\ -1 & -1 & e^{\lambda_{21}\cdot\tau_2} & e^{\lambda_{22}\cdot\tau_2} \\ e^{\lambda_{11}\cdot\tau_1} & e^{\lambda_{12}\cdot\tau_1} & -1 & -1 \end{pmatrix} \cdot \begin{pmatrix} A_{11} \\ A_{12} \\ A_{21} \\ A_{22} \end{pmatrix} = \begin{pmatrix} a_{\text{out}}^{(1)} \cdot q_{\text{out}} \\ a_{\text{out}}^{(2)} \cdot q_{\text{out}} \\ 0 \\ 0 \end{pmatrix}. \tag{3.35}$$

With Eq. 3.35, the coefficients A_{11}, A_{12}, A_{21}, and A_{22} can be calculated and inserted in Eq. 3.24. The current I_i can then be used in Eq. 2.14 for calculating the equivalent output resistance of a ReSC converter. Since the symbolic solution for R_{out} in the general case leads to a very complex equations, MATLAB® is used for solving the system of equations and the final R_{out} calculation.

Implementation of an Additional Conversion Ratio 4/7 by Three-Phase Operation

Figure 3.23 shows the different capacitor configurations for the three separate phases φ_1, φ_2, and φ_3 of the proposed 4/7 ratio power stage (see Fig. 3.4) along with the parameters from the charge flow analysis (see Sect. 2.3.1).

With Fig. 3.23, the charge multiplier vectors for the different phases can be determined

$$a_{4/7}^{(1)} = [4q_x \ 3q_x \ q_x \ q_x \ 4q_x]^{\text{T}} \cdot \frac{1}{7q_x} = \left[\frac{4}{7} \ \frac{3}{7} \ \frac{1}{7} \ \frac{1}{7} \ \frac{4}{7}\right]^{\text{T}} \tag{3.36}$$

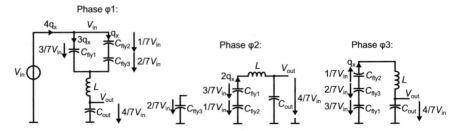

Fig. 3.23 Charge flow analysis for the 4/7 conversion of the resonant SC converter

$$a_{4/7}^{(2)} = [2q_x \ -2q_x \ -2q_x \ 0 \ 0]^T \cdot \frac{1}{7q_x} = \left[\frac{2}{7} \ -\frac{2}{7} \ -\frac{2}{7} \ 0 \ 0\right]^T \quad (3.37)$$

$$a_{4/7}^{(3)} = [q_x \ -q_x \ q_x \ -q_x \ 0]^T \cdot \frac{1}{7q_x} = \left[\frac{1}{7} \ -\frac{1}{7} \ \frac{1}{7} \ -\frac{1}{7} \ 0\right]^T \quad (3.38)$$

The ratio of the total input and output charge elements leads to the desired conversion ratio 4/7

$$N = \frac{\sum a_{in}^{(i)}}{\sum a_{out}^{(i)}} = \frac{\frac{4}{7}}{\frac{4}{7} + \frac{2}{7} + \frac{1}{7}} = \frac{4}{7}. \quad (3.39)$$

According to the charge flow analysis, the flying capacitor C_{fly3} delivers the smallest amount of charge. Therefore, the existing power stage (see Fig. 3.1) with $C_{fly1} = C_{fly2} = 1\,\text{nF}$ is extended by a small 250 pF capacitor for C_{fly3} as shown in Fig. 3.4b. Due to different voltage polarities across the power switches during operation, the power switches S2, S7, and S9 have to be implemented by a back-to-back transistor configuration.

Depending on the capacitor configuration in each phase, different resonance frequencies occur, which also lead to different duty cycles (see Sect. 3.1). This can be seen in Fig. 3.24, where the simulated transient waveforms of the inductor current I_L and the clock signals $\varphi1$, $\varphi2$, and $\varphi3$ are shown. The inductor current I_L in the individual phases matches the calculated output charge vectors ($a_{out,4/7}^{(1)} = 4/7$, $a_{out,4/7}^{(2)} = 2/7$, $a_{out,4/7}^{(3)} = 1/7$). The amount of charge is halved after each phase. The benefit of the 4/7 conversion ratio can be seen in Fig. 3.4a, where the simulated converter efficiency η is plotted over the input voltage V_{in}. The 4/7 ratio leads to an efficiency improvement of up to 7%.

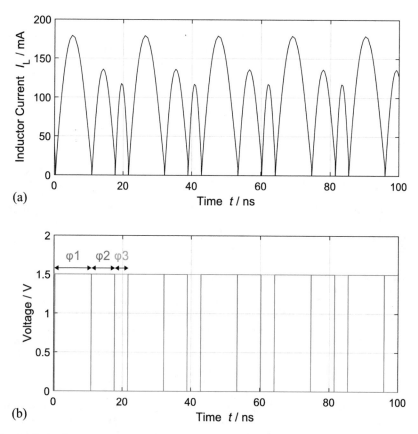

Fig. 3.24 Simulated transient waveforms of the 4/7 conversion ratio for $V_{in} = 3.4$ V, $V_{out} = 1.8$ V, $I_{out} = 100$ mA: (**a**) inductor current I_L; (**b**) clock signals $\varphi 1$, $\varphi 2$, and $\varphi 3$

References

1. Le, H., Sanders, S.R., Alon, E.: Design techniques for fully integrated switched-capacitor DCDC converters. IEEE J. Solid-State Circuits **46**(9), 2120–2131 (2011). ISSN: 1558-173X. https://doi.org/10.1109/JSSC.2011.2159054
2. Piqué, G.V.: A 41-phase switched-capacitor power converter with 3.8mV output ripple and 81% efficiency in baseline 90nm CMOS. In: 2012 IEEE International Solid-State Circuits Conference, pp. 98–100 (2012). https://doi.org/10.1109/ISSCC.2012.6176892
3. Andersen, T.M., et al.: A sub-ns response on-chip switched-capacitor DCDC voltage regulator delivering 3.7W/mm2 at 90% efficiency using deep-trench capacitors in 32nm SOI CMOS. In: 2014 IEEE International Solid-State Circuits Conference Digest of Technical Papers (ISSCC), pp. 90–91 (2014). https://doi.org/10.1109/ISSCC.2014.6757351
4. Salem, L.G., Mercier, P.P.: An 85%-efficiency fully integrated 15-ratio recursive switched-capacitor DCDC converter with 0.1-to-2.2V output voltage range. In: 2014 IEEE International Solid-State Circuits Conference Digest of Technical Papers (ISSCC), pp. 88–89 (2014). https://doi.org/10.1109/ISSCC.2014.6757350

5. Lutz, D., Renz, P., Wicht, B.: A 10mW fully integrated 2- to-13V-input buck-boost SC converter with 81.5% peak efficiency. In: 2016 IEEE International Solid-State Circuits Conference (ISSCC), pp. 224–225 (2016a). https://doi.org/10.1109/ISSCC.2016.7417988

6. Sarafianos, A., et al.: A folding Dickson-based fully integrated wide input range capacitive DCDC converter achieving Vout/2-resolution and 71% average efficiency. In: 2015 IEEE Asian Solid-State Circuits Conference (A-SSCC), pp. 1–4 (2015). https://doi.org/10.1109/ASSCC. 2015.7387488

7. Jiang, J., et al.: A 2-/3-phase fully integrated switched-capacitor DCDC converter in bulk CMOS for energy-efficient digital circuits with 14% efficiency improvement. In: 2015 IEEE International Solid-State Circuits Conference – (ISSCC) Digest of Technical Papers, pp. 1–3 (2015). https://doi.org/10.1109/ISSCC.2015.7063078

8. Makowski, M.S.: On performance limits of switched-capacitor multi-phase charge pump circuits. Remarks on papers of Starzyk et al. In: 2008 International Conference on Signals and Electronic Systems, pp. 309–312 (2008). https://doi.org/10.1109/ICSES.2008.4673422

9. Starzyk, J.A., Jan, Y.-W., Qiu, F.: A DCDC charge pump design based on voltage doublers. In: IEEE Trans. Circuits Syst. I: Fundamental Theory Appl. **48**(3), 350–359 (2001). ISSN: 1558-1268. https://doi.org/10.1109/81.915390

10. Karadi, R., Pique, G.V.: 3-phase 6/1 switched-capacitor DCDC boost converter providing 16V at 7mA and 70.3% efficiency in 1.1mm3. In: 2014 IEEE International Solid-State Circuits Conference Digest of Technical Papers (ISSCC), pp. 92–93 (2014). https://doi.org/10.1109/ISSCC.2014.6757352

11. Bang, S., Blaauw, D., Sylvester, D.: A successive-approximation switched-capacitor DC–DC converter with resolution of $V_{IN}/2^N$ for a wide range of input and output voltages. IEEE J. Solid-State Circuits **51**(2), 543–556 (2016). https://doi.org/10.1109/JSSC.2015.2501985

12. Nguyen, B., et al.: High-efficiency fully integrated switched-capacitor voltage regulator for battery-connected applications in low-breakdown process technologies. IEEE Trans. Power Electron. **33**(8), 6858–6868 (2018). https://doi.org/10.1109/TPEL.2017.2757950

13. Lutz, D., Renz, P., andWicht, B.: A 120/230Vrms-to-3.3V micro power supply with a fully integrated 17V SC DCDC converter. In: ESSCIRC Conference 2016: 42nd European Solid-State Circuits Conference, pp. 449–452 (2016b). https://doi.org/10.1109/ESSCIRC.2016.7598338

14. Lutz, D., Renz, P., Wicht, B.: An integrated 3-mW 120/230-V AC mains micropower supply. IEEE J. Emerg. Sel. Top. Power Electron. **6**(2), 581–591 (2018). ISSN: 2168-6777. https://doi. org/10.1109/JESTPE.2018.2798504

15. Cheng, K.W.E.: New generation of switched capacitor converters. In: PESC 98 Record. 29th Annual IEEE Power Electronics Specialists Conference (Cat. No.98CH36196), vol. 2, pp. 1529–1535 (1998). https://doi.org/10.1109/PESC.1998.703377

16. Lin, Y.-C., Liaw, D.-C.: Parametric study of a resonant switched capacitor DCDC converter. In: Proceedings of IEEE Region 10 International Conference on Electrical and Electronic Technology. TENCON 2001 (Cat. No.01CH37239), vol. 2, pp. 710–716 (2001). https://doi. org/10.1109/TENCON.2001.949684

17. Kesarwani, K., Sangwan, R., Stauth, J.T.: A 2-phase resonant switched-capacitor converter delivering 4.3W at 0.6W/mm2 with 85% efficiency. In: 2014 IEEE International Solid-State Circuits Conference Digest of Technical Papers (ISSCC), pp. 86–87 (2014). https://doi.org/10. 1109/ISSCC.2014.6757349

18. Kesarwani, K., Stauth, J.T.: Resonant and multi-mode operation of flying capacitor multi-level DCDC converters. In: 2015 IEEE 16th Workshop on Control and Modeling for Power Electronics (COMPEL), pp. 1–8 (2015). https://doi.org/10.1109/COMPEL.2015.7236511

19. Renz, P., Lueders, M., Wicht, B.: A 47 MHz Hybrid Resonant SC Converter with Digital Switch Conductance Regulation and Multi-Mode Control for Li-Ion Battery Applications, 2020 IEEE Applied Power Electronics Conference and Exposition (APEC), New Orleans, LA, USA, 2020, pp. 15–18, https://doi.org/10.1109/APEC39645.2020.9124238.

20. Ng, V.W., Sanders, S.R.: A high-efficiency wide-input- voltage range switched capacitor point-of-load DCDC converter. IEEE Trans. Power Electron. **28**, 4335–4341 (2013)

21. Pasternak, S.R., et al.: Modeling and performance limits of switched-capacitor DC–DC converters capable of resonant operation with a single inductor. IEEE J. Emerg. Sel. Top. Power Electron. **5**(4), 1746–1760 (2017). ISSN: 2168-6785. https://doi.org/10.1109/JESTPE. 2017.2730823

22. Seeman, M.D.: A design methodology for switched-capacitor DCDC converters. PhD thesis. EECS Department, University of California, Berkeley (2009). https://www2.eecs.berkeley.edu/Pubs/TechRpts/2009/EECS-2009-78.html

23. Schaef, C., Kesarwani, K., Stauth, J.T.: A variable-conversion-ratio 3-phase resonant switched capacitor converter with 85% efficiency at 0.91W/mm2 using 1.1nH PCB-trace inductors. In: 2016 IEEE International Solid-State Circuits Conference – (ISSCC) Digest of Technical Papers, pp. 1–3 (2015). https://doi.org/10.1109/ISSCC.2015.7063075

24. McLaughlin, P.H., Xia, Z., Stauth, J.T.: A fully integrated resonant switched-capacitor converter with 85.5% efficiency at 0.47W using on-chip dual-phase merged-LC resonator. In: 2020 IEEE International Solid-State Circuits Conference – (ISSCC), pp. 192–194 (2020)

25. Krishnamurthy, H.K., et al.: A digitally controlled fully integrated voltage regulator with on-die solenoid inductor with planar magnetic core in 14-nm tri-gate CMOS. IEEE J. Solid-State Circuits **53**(1), 8–19 (2018). ISSN: 0018-9200. https://doi.org/10.1109/JSSC.2017.2759117

26. Amin, S.S., Mercier, P.P.: A fully integrated Li-ion-compatible hybrid four-level DCDC converter in 28-nm FDSOI. IEEE J. Solid-State Circuits **54**(3), 720–732 (2019). ISSN: 0018-9200. https://doi.org/10.1109/JSSC.2018.2880183

27. Godycki, W., Sun, B., Apsel, A.: Part-time resonant switching for light load efficiency improvement of a 3-level fully integrated buck converter. In: ESSCIRC 2014 – 40th European Solid State Circuits Conference (ESSCIRC), pp. 163–166 (2014). https://doi.org/10.1109/ESSCIRC.2014.6942047

28. Kim, W., Brooks, D., Wei, G.: A fully-integrated 3-levelDCDC converter for nanosecond-scale DVFS. IEEE J. Solid-State Circuits **47**(1), 206–219 (2012). ISSN: 0018-9200. https://doi.org/10.1109/JSSC.2011.2169309

29. Kumar, P., et al.: A 0.4V-1V 0.2A/mm2 70% efficient 500MHz fully integrated digitally controlled 3-level buck voltage regulator with on-die high density MIM capacitor in 22nm tri-gate CMOS. In: 2015 IEEE Custom Integrated Circuits Conference (CICC), pp. 1–4 (2015). https://doi.org/10.1109/CICC.2015.7338479

30. Liu, W., et al.: A 94.2%-peak-efficiency 1.53A direct-battery-hook-up hybrid Dickson switched-capacitor DCDC converter with wide continuous conversion ratio in 65nm CMOS. In: 2017 IEEE International Solid-State Circuits Conference (ISSCC), pp. 182–183 (2017). https://doi.org/10.1109/ISSCC.2017.7870321

31. Li, Y., et al.: AC-coupled stacked dual-active-bridge DC–DC converter for integrated lithium-ion battery power delivery. IEEE J. Solid-State Circuits **54**(3), 733–744 (2019). ISSN: 1558-173X. https://doi.org/10.1109/JSSC.2018.2883746

32. Renz, P., et al.: A fully integrated 85%-peak-efficiency hybrid multi ratio resonant DCDC converter with 3.0-to-4.5V input and 500μA -to- 120mA load range. In: 2019 IEEE International Solid-State Circuits Conference – (ISSCC), pp. 156–158 (2019a). https://doi.org/10.1109/ISSCC.2019.8662491

33. Renz, P., '.: A 3-ratio 85% efficient resonant SC converter with on-chip coil for Li-ion battery operation. IEEE Solid-State Circuits Lett. **2**(11), 236–239 (2019b). ISSN: 2573-9603. https://doi.org/10.1109/LSSC.2019.2927131

34. Renz, P., Deneke, N., Wicht, B.: Dynamic Modeling and Control of a Resonant Switched-Capacitor Converter with Switch Conductance Regulation, 2020 IEEE 21st Workshop on Control and Modeling for Power Electronics (COMPEL), Aalborg, Denmark, 2020, pp. 1–4, https://doi.org/10.1109/COMPEL49091.2020.9265644.

Chapter 4
Circuit Implementation for High Efficiency

One of the key challenges for high efficient converter operation is the implementation and the control of the power switches. The required low-power circuit blocks for reliable and high efficient operation are presented and discussed in this chapter.

In Sect. 4.1, the efficiency benefit of low-voltage transistor stacking over single high-voltage switches is investigated. The segmentation of the power switches, required for switch conductance regulation, is shown in Sect. 4.2. The implementation of different level shifters for the control of the power switches is discussed in Sect. 4.3, while Sect. 4.4 covers the generation of a flying supply voltage for each power switch. Section 4.5 presents the complete power stage implementation.

4.1 Transistor Stacking

Stacking of low-voltage transistors as outlined in Fig. 4.1 is widely used in different applications like line drivers, power amplifiers, or power management circuits [1–14]. In low-voltage deep sub-micrometer CMOS technologies, stacking of low-voltage transistors is often done because of a lack of available high-voltage devices [2–13]. Stacked low-voltage transistors can also be beneficial in terms of efficiency even if high-voltage devices are available [15]. Especially in power management circuits designed in larger process nodes, the designer can choose between a stacked option with thin-oxide low-voltage devices and a single high-voltage transistor.

The principle of transistor stacking is shown in Fig. 4.1. For the sake of simplicity, the maximum voltage across the terminals of a transistor is limited to nominal voltage V_{DD} of the core devices in the used process in the following explanations (see Fig. 4.1a). Stacking of multiple transistors leads to a higher blocking capability of $n \cdot V_{DD}$ (Fig. 4.1b). In such a configuration, proper gate biasing of the transistors is essential. During operation, each transistor has to be biased in a way that the voltage across its terminals remains within the nominal voltage limit.

© The Editor(s) (if applicable) and The Author(s), under exclusive license to
Springer Nature Switzerland AG 2021
P. Renz, B. Wicht, *Integrated Hybrid Resonant DCDC Converters*,
https://doi.org/10.1007/978-3-030-63944-0_4

Fig. 4.1 Principle of
low-voltage transistor
stacking: (**a**) Maximum
allowed voltage across the
terminals of a low-voltage
transistor; (**b**) Stacking of n
low-voltage transistors

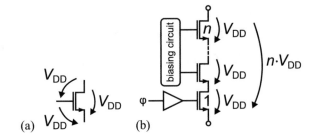

Previous implementations of stacked low-voltage transistors were mostly limited
to simple half-bridges for I/O drivers or inductive buck converters [1–14]. The trend
in DCDC converters continues toward hybrid converters with more elaborate power
stages [16–21] and higher switching frequencies in order to reduce the size of
the passive storage components. In these published topologies as well as for the
proposed power stage (see Sect. 3.1), the implementation of efficient power switches
gets more challenging since there are more switches and different flying voltage rails
involved. Appropriate low-voltage transistor stacking can improve the converter
efficiency significantly, but it has to be capable to work with arbitrary flying voltage
levels.

Section 4.1.1 gives an overview of different implementation options for the
power switches. For a fair comparison of the losses, the different power switch
options are modeled in Sect. 4.1.2. A detailed loss comparison is given in Sect. 4.1.3.

4.1.1   Overview of Different Implementation Options for Power Switches

The blocking voltage V_{blk} is the fundamental criterion for the power switch. Simple
half-bridges, which operate at higher input voltages, often have to block a maximum
input voltage, e.g., the Li-Ion battery ranges up to $V_{blk} = V_{in} = 4.5\,\text{V}$. Advanced
DCDC converter topologies, such as [16–23], use more elaborate power stages
where the power switches have to block only a fraction of the input voltage. There
are several criteria for the choice of one transistor type as a power switch. This
section compares different solutions for the implementation of power switches with
respect to switching and conduction losses and blocking voltage capability.

Single 5 V Transistor Implementation

The straightforward option to implement a single power switch is the use of thick-
oxide NMOSand PMOStransistors with 5 V rating. They can be driven with the
available potentials, the blocking voltage V_{blk} and the global reference potential

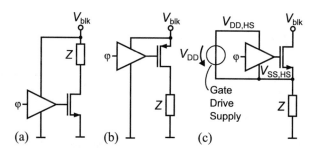

Fig. 4.2 Different options for single 5 V transistor implementation: (**a**) low-side NMOS transistor directly driven with V_{blk} and GND; (**b**) high-side PMOS transistor directly driven with V_{blk} and GND; (**c**) high-side NMOS transistor driven with floating gate drive supply

GND shown in Fig. 4.2a,b with a NMOS transistor as a low-side switch and PMOS device as a high-side transistor. These options generate nearly no design overhead, but they are limited to applications where the blocking voltage V_{blk} is lower than the maximum gate-source voltage $V_{GS,max}$. It works well for switches that are referred to one of the voltage rails (V_{blk} or GND respectively). This is the case in a half-bridge used for conventional buck converters, line drivers, or power amplifiers. If the switches are referred to different intermediate voltage rails, V_{blk}- or GND-driven transistors can be used in some cases, but the reduced overdrive voltage leads to larger area transistors to provide the same on-resistance $R_{DS,on}$. At higher blocking voltages $V_{blk} > V_{GS,max}$, single NMOS or PMOS device driven by V_{blk}/GND cannot be used. In such cases, a transistor switch requires a floating gate drive supply (Fig. 4.2c). The use of a NMOS transistor is preferable as it results in lower area and on-resistance than a PMOS transistor.

Stacking of Two 1.5 V Transistors

A stack of two 1.5 V thin-oxide transistors provides sufficient blocking capability of $V_{blk} = 2V_{DD}$ with the nominal voltage V_{DD} of the core devices in the particular process. An easy and commonly used way for stacking two low-voltage transistors is shown in Fig. 4.3, implemented in designs like in [3, 4, 14, 22, 23]. The lower transistor M1 is directly driven by the driver circuit, while the upper transistor M2 is biased at V_{DD} from the gate drive supply. While transistors M1 and M2 are turned on, they operate in the triode region, and the node V_1 discharges to $V_{DD,HS}$. When the clock signal φ turns low, transistor M1 turns off. The potential at node V_1 rises to $V_1 = V_{DD,HS} - V_{th}$ and may then slowly rise to $V_{DD,HS}$ due to the subthreshold current of M2. This way, transistor M2 cuts off as well. At the beginning of the turn-off transient, M2 sees a drain-source voltage V_{DS2} larger than $V_{DD,HS}$. For $V_{blk} = 2V_{DD}$ and $V_1 - V_{SS,HS} = V_{DD} - V_{th}$, the loop calculation results in $V_{DD} + V_{th}$. With a typical maximum rating for the drain-source voltage of $V_{DS,max} \sim 1.5V_{DD}$, this results in $V_{DD} > 2V_{th}$ as a condition for reliable operation, which is well

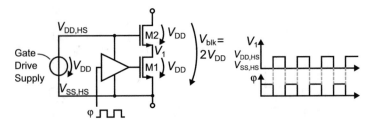

Fig. 4.3 Stacking of two 1.5 V transistors with a total blocking capability of $V_{blk} = 2V_{DD}$

fulfilled in this work. While this is inherently obtained in designs with transistors of higher-voltage classes (higher blocking voltages), it has to be considered for switch stacking in advanced technologies with very low V_{DD} values. At the clock signal transition from low to high, first transistor M1 discharges the node V_1 to the reference potential $V_{SS,HS}$ of this switch, which can be the global reference potential GND or any arbitrary voltage. Now, transistor M2 turns on. For all switches not referred to global GND potential, a gate drive supply is required which generates the gate-overdrive voltage $V_{DD,HS}$ on the flying high-side domain (see Sect. 4.4). Section 4.1.3 shows that stacking of two thin-oxide transistors leads to significantly lower losses compared to a single thick-oxide transistor implementation.

Stacking of Three 1.5 V Transistors

If higher-voltage blocking capability is required, the transistor stack can be extended to three low-voltage devices, Fig. 4.4. In [13], three stacked transistors are implemented, but they can only withstand a blocking voltage of $V_{blk} = 2.5 \cdot V_{DD}$. This is due to the fact that not all of the stacked transistors are switched from triode region to cut-off region and vice versa. Instead, one switch operates as a MOS-diode leading to lower blocking voltage.

Figure 4.4 shows the implementation of three stacked low-voltage transistors for the power switch implementation, which can withstand a blocking voltage of $V_{blk} = 3 \cdot V_{DD}$. Figure 4.4a is often used in half-bridges of buck converters [1, 3, 4] and uses an additional static bias rail V_{bias1}, a level shifter, and a driver for the third transistor M3. Transistors M1 and M2 work in the same way as described for a two-transistor stack. Transistor M3 is controlled with a separate level shifter and gate driver circuit, which is supplied by the two voltage rails V_{DD} and $V_{bias1} = 2 \cdot V_{DD}$. The additional bias rail V_{bias1} is often supplied externally [1, 3, 4]. In this case only fixed input voltages can be supported. The complex generation of the additional voltage rail is only included in [24].

The option in Fig. 4.4b does not need an additional level shifter and driver circuit [8–11]. The bias voltage for the third transistor M3 is generated with the switches M4 and M5 and resistive dividers R_1-R_2 and R_3-R_4. In off-state, the voltage on V_{G3} is provided by the resistive divider (R_1-R_2) between the drain of M3 and the fixed

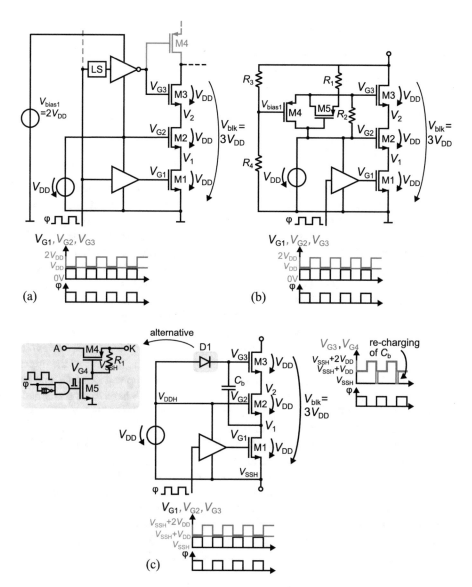

Fig. 4.4 Different options for stacking of three low-voltage transistors for blocking capability of $V_{blk} = 3V_{DD}$: (**a**) [1, 3, 4]; (**b**) [10]; (**c**) this work

bias voltage of V_{G2}. This is controlled by M5, which operates in the triode region. M4 is cut-off, set by R_3 and R_4. In the on-state, the divider R_1-R_2 is disabled by M5, and the gate voltage V_{G3} is supplied by the fixed bias voltage of M2 which passes through M4. A static current flow through the resistive dividers and the dependency on the absolute blocking voltage across the transistor stack are major limitations of this solution.

Table 4.1 Comparison of different options for stacking of three low-voltage transistors

	[1, 3] Fig. 4.4a	[10] Fig. 4.4b	**This work** Fig. 4.4c
Needs additional voltage supplies	Yes	No	No
Static power consumption	N.a.	Yes	No
Needs additional level shifter, driver	Yes	No	No
Symmetrical voltage distribution	Yes	Yes	Yes
Max. blocking voltage V_{blk}	$3V_{DD}$	$3V_{DD}$	$3V_{DD}$

As part of this work, the solution in Fig. 4.4c is proposed. It enables lossless biasing of transistor M3 and supports arbitrary flying voltage levels at node $V_{SS,HS}$. No additional level shifters and drivers are necessary. The correct gate voltage V_{G3} is generated with a small bootstrap capacitance C_b, which is referred to node V_1. With this bootstrap circuit, the voltage V_{G3} automatically switches between $V_{SS,HS} + 2V_{DD}$ and $V_{SS,HS} + V_{DD}$. The bootstrap capacitance C_b is recharged from the gate drive supply via the diode D1 during the on-phase of the clock signal φ. For a small voltage drop, a Schottky diode should be used. If no Schottky diode is available, the bootstrap circuit formed by the transistors M4 and M5 and resistor R_1 can be used. It allows for charging of the bootstrap capacitance C_b without a diode voltage drop. Transistor M4 is turned on with a small pulse by transistor M5 in order to avoid static current flow. Since there is a charge backflow during the turn-off of the upper transistor M3, the pulse width for recharging of the bootstrap capacitor C_b can be short since only gate leakage and charge redistribution losses have to be recharged. The bootstrap capacitor C_b can be small and is designed to be seven times greater than the gate capacitance of transistor M3.

Table 4.1 shows a comparison of the different options of Fig. 4.4. Only the proposed circuit of Fig. 4.4c supports arbitrary voltage levels at $V_{SS,HS}$ together with no static power consumption and no additional level shifters and drivers. Moreover, Sect. 4.1.3 shows that the proposed stacking of three thin-oxide transistors leads to significantly lower losses compared with a single thick-oxide transistor implementation. The proposed approach can also be extended to stacking of more than three transistors for even higher blocking voltages. An example of four stacked devices with a blocking capability $V_{blk} = 4V_{DD}$ is given in Fig. 4.5.

4.1.2 Modeling of Different Power Switch Options

The total power dissipation consists of a static and a dynamic part. Conduction losses P_{cond} are dissipated in the on-resistance of the switch. The charging and discharging of parasitic capacitors of the transistors correspond to dynamic switching losses P_{sw}. Transition losses during the switching event are neglected for the comparisons. It can be assumed that they are the same for all switch options.

Fig. 4.5 Extension of
Fig. 4.4c to four stacked
low-voltage transistors with a
blocking capability of
$V_{blk} = 4V_{DD}$

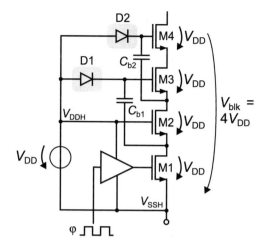

The total losses can thus be calculated as

$$P_{tot} = P_{sw} + P_{cond}. \tag{4.1}$$

For a single 5 V NMOS/PMOS transistor, the switching and conduction losses can be calculated by

$$P_{sw,5V} = (C_{GS}V_{GS}^2 + C_{GD}(V_{GS} + V_{blk})^2)f_{sw} \cdot (1/\eta_{CP}) \tag{4.2}$$

$$P_{cond,5V} = R_{DS,on} \cdot I_{RMS}^2 \tag{4.3}$$

where I_{RMS} is the RMS current through the transistor and η_{CP} is the efficiency of the charge pump. If the transistors are driven directly with the blocking voltage V_{blk} (Fig. 4.2a,b), no charge pump is required for gate overdrive generation, which results in $\eta_{CP} = 1$. In this case, the gate-source voltage V_{GS} depends on the application and cannot be chosen arbitrarily. With a floating gate drive supply (Fig. 4.2c), the gate-source voltage V_{GS} of the 5 V transistors could be chosen arbitrarily in the range between $V_{GS} = V_{th}$ and $V_{DD} = 5$ V. In order to evaluate the impact of the turn-on voltage on the overall losses and to find a minimum $E_{loss,min}$ for the total losses, the resistance $R_{DS,on}$ and the gate-source voltage V_{GS} are swept as shown in Fig. 4.6 for a typical duty cycle of 50%. Close to the minimum, the total energy loss remains nearly constant and is independent of the gate-source voltage for $V_{GS} > 2$ V. A lower gate-source voltage V_{GS} does not lead to lower capacitive switching losses because it also leads to larger transistor area for same $R_{DS,on}$. Thus, higher gate capacitance values cancel the efficiency benefit of a lower turn-on voltage V_{GS}. Especially the gate-drain capacitance C_{GD} gets dominant since the blocking voltage V_{blk} remains the same.

Fig. 4.6 Total energy losses
E_{loss} of 5 V NMOS transistor
vs. on-resistance $R_{\mathrm{DS,on}}$ and
gate-source voltage V_{GS} (at
$I_{\mathrm{RMS}} = 100\,\mathrm{mA}$)

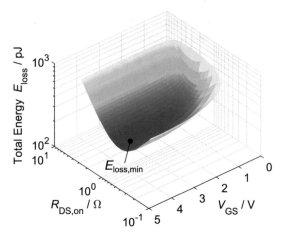

For the calculation of the losses of two and three stacked transistors, the
equivalent circuit depicted in Fig. 4.7 is used. It shows all relevant capacitors
and voltage swings (indicated by ↔) at each terminal. Since the impact of the
gate-source voltage on the total losses is low (see Fig. 4.6), a turn-on voltage of
$V_{\mathrm{GS}} = V_{\mathrm{DD}} = 1.5\,\mathrm{V}$ was chosen for an area-efficient implementation. The gray
marked area at the top of both parts of Fig. 4.7 has to be considered for three stacked
transistors according to the proposed solution of Fig. 4.4c. The total losses for both
two and three stacked transistors can be calculated from (4.1). These consist of
the dynamic switching losses $P_{\mathrm{sw,CP}}$ delivered by the charge pump, the switching
losses P_{sw} delivered by the power stage itself, and the conduction losses P_{cond}. For
two stacked transistors, they can be calculated by

$$P_{\mathrm{sw,CP,2stack}} = \left(C_{\mathrm{GS1}} V_{\mathrm{DD}}^{\,2} + C_{\mathrm{GD1}} (2V_{\mathrm{DD}} - V_{\mathrm{th}})^2 \right) f_{\mathrm{sw}} \cdot (1/\eta_{\mathrm{CP}}) \tag{4.4}$$

$$P_{\mathrm{sw,2stack}} = \Big(C_{\mathrm{DS1}} (V_{\mathrm{DD}} - V_{\mathrm{th}})^2 + C_{\mathrm{GS2}} (V_{\mathrm{DD}} - V_{\mathrm{th}})^2$$
$$+ C_{\mathrm{GD2}} V_{\mathrm{blk},2}^{\,2} + C_{\mathrm{DS2}} (V_{\mathrm{blk},2} - (V_{\mathrm{DD}} - V_{\mathrm{th}}))^2 \Big) f_{\mathrm{sw}} \tag{4.5}$$

$$P_{\mathrm{cond,2stack}} = 2 R_{\mathrm{DS,on}} \cdot I_{\mathrm{RMS}}^{\,2}. \tag{4.6}$$

For three stacked transistor, the gray marked area in Fig. 4.7 has to be considered
additionally, which leads to the following equations:

$$P_{\mathrm{sw,CP,3stack}} = \left(C_{\mathrm{GS1}} V_{\mathrm{DD}}^{\,2} + C_{\mathrm{GD1}} (2V_{\mathrm{DD}} - V_{\mathrm{th}})^2 \right) f_{\mathrm{sw}} \cdot (1/\eta_{\mathrm{CP}}) \tag{4.7}$$

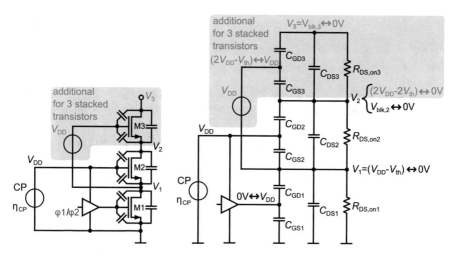

Fig. 4.7 Equivalent circuit for loss calculation of two and three stacked transistors

$$P_{\text{sw,3stack}} = \Big(C_{\text{DS1}}(V_{\text{DD}} - V_{\text{th}})^2 + C_{\text{GS2}}(V_{\text{DD}} - V_{\text{th}})^2$$
$$+ C_{\text{GD2}}(2V_{\text{DD}} - 2V_{\text{th}})^2 + C_{\text{DS2}}(V_{\text{DD}} - V_{\text{th}})^2 + C_{\text{GS3}}(V_{\text{DD}})^2$$
$$+ C_{\text{GD3}}(V_{\text{blk,3}} - V_{\text{DD}} + V_{\text{th}})^2 + C_{\text{DS3}}(V_{\text{blk,3}} - (2V_{\text{DD}} - V_{\text{th}}))^2\Big) f_{\text{sw}}$$

$$(4.8)$$

$$P_{\text{cond,3stack}} = 3R_{\text{DS,on}} \cdot I_{\text{RMS}}^2 \qquad (4.9)$$

Equations 4.4 to 4.9 assume an equal on-resistance $R_{\text{DS,on1}} = R_{\text{DS,on2}} = R_{\text{DS,on3}} = R_{\text{DS,on}}$. $V_{\text{blk,2}}$ and $V_{\text{blk,3}}$ are the corresponding blocking voltage for the two and three stacked devices, which occur in the application. As an advantage of this configuration, only the parasitic capacitors C_{GS1} and C_{GD1} of the transistor M1 have to be charged by the charge pump with its efficiency η_{CP} ($P_{\text{sw,CP}}$) for both options (see Eqs. 4.4 and 4.7). This brings significant loss reduction because all other parasitic capacitances are directly charged from input (V_{blk}) without any additional losses (see Eqs. 4.5 and 4.8). They are denoted by P_{sw}.

The loss distribution for three stacked transistors is shown in Fig. 4.8 at the minimum of the total losses E_{loss} (see Fig. 4.12a). Therefore, the total losses split equally between the switching losses ($P_{\text{sw,CP}} + P_{\text{sw}}$) and the conduction losses P_{cond}. A typical efficiency of $\eta_{\text{CP}} = 50\%$ is assumed (see Sect. 4.4), which is already included in the switching loss. The loss distribution for two stacked transistors is similar to Fig. 4.8.

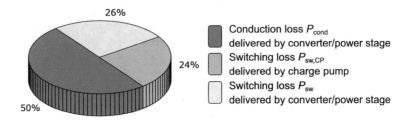

Fig. 4.8 Loss distribution for three stacked transistors at the minimum of total loss E_{loss}

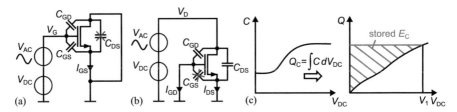

Fig. 4.9 Extraction of parasitic capacitors: (**a**) circuit for extraction of C_{GS} capacitance; (**b**) circuit for extraction of C_{GD} and C_{DS} capacitance; (**c**) extraction of the $C = f(V_{DC})$ and $Q = f(V_{DC})$ curves from simulation

Extraction of Energy Equivalent Transistor Capacitances

Prior work often uses the conventional oxide capacitance expression, which does not distinguish between gate-source and gate-drain capacitance [25–27]. Hence, a loss comparison has limited accuracy. Instead, as part of this work, an energy-based equivalent capacitance is used. In order to be able to apply Eqs. 4.4 to 4.9, an energy-based equivalent capacitance value C_{eq} for the capacitances C_{GS}, C_{GD}, and C_{DS} is extracted from simulations.

First, the nonlinear capacitances $C = f(V_{DC})$ are derived from the test circuits in Fig. 4.9a. These can now be used to calculate the charge $Q_C = f(V_{DC})$ according to Fig. 4.9b. The stored energy E_C can be calculated from (4.10)

$$E_C(V_1) = \int V_{DC} \, dQ_C$$

$$= Q_C(V_1) \cdot V_1 - \int_0^{V_1} Q_C(V_{DC}) \, dV_{DC} \qquad (4.10)$$

E_C allows to calculate the energy-based equivalent capacitance C_{eq} according to (4.11)

$$C_{eq} = \frac{2 \cdot E_C(V_1)}{V_1^2} \qquad (4.11)$$

This energy-based equivalent capacitance value C_{eq} is precise at the voltages V_{DC}. To be safe and to ensure a fair comparison among the different options, the maximum occurring voltage during operation for each transistor type is used in the extraction. In order to calculate the total losses of the different transistor options, the equivalent capacitance values can be inserted into Eqs. 4.4, 4.5, 4.7, and 4.8 for all parasitic transistor capacitances, i.e., C_{GS1}, C_{DS1}, etc., in order to calculate the total losses of the different transistor options.

4.1.3 Experimental Results

Different options for the power switches are fabricated in a 130 nm BCD process with 1.5 V core devices along with 5 V high-voltage switches. Test structures with single 5 V NMOS/PMOS transistors (Fig. 4.2a,b), two stacked 1.5 V transistors (Fig. 4.3), and three stacked 1.5 V transistors (Fig. 4.4c) are implemented for comparison of the different options. For the implementation of the power stage, a stack of two low-voltage 1.5 V transistors is used since it has sufficient voltage blocking capability as described in Sect. 4.5.1. The power switch consists of the transistor stack together with the gate driver (see Sect. 4.5.2) and different level shifters (see Sect. 4.3). It also includes a charge pump for gate overdrive generation which is introduced in Sect. 4.4.

Transient Measurements

Figure 4.10 shows transient measurements of the voltages at the terminals for the different stacked transistor options. The node voltage V_1 of the two-1.5 V-transistor stack in Fig. 4.10a during turn-off is exactly half of the blocking voltage $V_{blk} = 3$ V. This results in drain-source voltages of transistors M1 and M2 of around $V_{DS1} = V_{DS2} = 1.5$ V. In Fig. 4.10b the measured waveforms for the three stacked 1.5 V transistors for a blocking voltage of $V_{blk} = 4.5$ V are shown. In the blocking state, the node voltages are approximately $V_1 = 1.5$ V and $V_2 = 3$ V, respectively. This leads to a symmetrical distribution of the drain-source voltages of $V_{DS1} = V_{DS2} = V_{DS3} = 1.5$ V. For the measurements, a resistive load is used. Inductive loads and various parasitics may cause ringing and voltage overshoot across the switches. As in any application with single switches, this has to be considered with respect to the maximum blocking capability of the switches.

Model vs. Measurement Results

Figure 4.11 shows a comparison between the loss model and measurements results, where the total capacitive switching losses P_{sw} for each option are plotted over the switching frequency f_{sw} for a duty cycle of 50%. All different options were

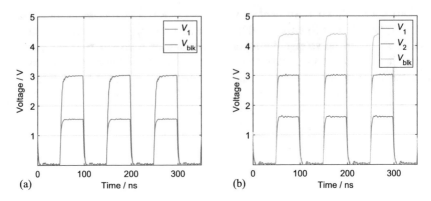

Fig. 4.10 Transient measurements of the voltages on the terminals of the stacked transistors: **(a)** two stacked 1.5 V transistors at $V_{\mathrm{blk}} = 3$ V; **(b)** three stacked 1.5 V transistors at $V_{\mathrm{blk}} = 4.5$ V

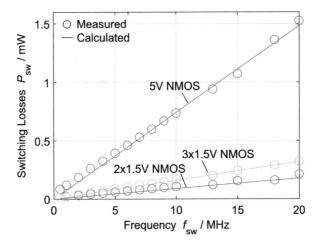

Fig. 4.11 Switching losses P_{sw} of the different options vs. switching frequency f_{sw} based on measurements and the presented loss model

implemented with the same on-resistance $R_{\mathrm{DS,on}} = 500\,\mathrm{m\Omega}$. The calculated values from the loss model track the measurements with less than 3% error. For the same on-resitance $R_{\mathrm{DS,on}}$, stacking of two or three 1.5 V transistors has a factor of 4.2 or 7.5 lower switching losses if compared to a 5 V NMOSimplementation. This confirms that switch stacking brings a significant efficiency benefit.

Comparison Between the Different Options

Based on the loss model (according to Eqs. 4.4 to 4.9), Fig. 4.12a shows the total calculated energy losses E_{loss} versus the on-resistance $R_{\mathrm{DS,on}}$ for the different options (Figs. 4.2, 4.3, and 4.4c) of the power switches. Depending on the appli-

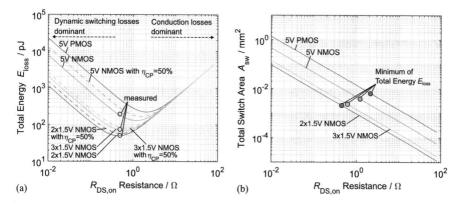

Fig. 4.12 (a) Total energy loss E_{loss} versus on-resistance $R_{DS,on}$ for the 5 V NMOS/PMOS transistor options (at $V_{blk} = 4.5$ V), two stacked 1.5 V transistors (at $V_{blk} = 3$ V), and three stacked 1.5 V transistors (at $V_{blk} = 4.5$ V and $I_{RMS} = 40$ mA); (b) total switch area A_{sw} versus $R_{DS,on}$ for different transistor options

cation (low-side or high-side switch), different curves have to be compared. For a fair comparison, the respective minimum of the total losses, which occurs at different $R_{DS,on}$ values, is considered. Equations 4.3, 4.6, and 4.9 show that the conduction losses are proportional to I_{RMS} while the switching losses do not depend on I_{RMS}. Therefore, the minimum and corresponding resistance value $R_{DS,on}$ will vary with I_{RMS}. However the relative comparison stays the same. For this reason, Fig. 4.12a is representative for the various switch stacking options. Even though the two stacked 1.5 V transistor option requires a dedicated charge pump for gate overdrive generation, it has factor of 2.3 or 3.9 lower energy loss E_{loss} compared to a 5 V NMOS or PMOS switch, respectively. If no charge pump is required (e.g., for a low-side switch), the total energy loss E_{loss} is 2.6 or 4.5 times lower compared to the 5 V NMOS or PMOS options. The three stacked 1.5 V transistors can withstand blocking voltages of up to $V_{blk} = 4.5$ V and have a factor 1.7 or 2.9 lower total losses E_{loss} (1.9 or 3.3 lower if no charge pump is required) compared with the 5 V NMOS or PMOS option. These results show that stacking of low-voltage transistors is the most efficient solution even with additional circuits such as a charge pump for gate overdrive generation. Figure 4.12a also indicates that the measurements (marked with circles) match the calculated values from the model.

Figure 4.12b shows the total switch area A_{sw} versus the on-resistance $R_{DS,on}$ for the different implementation options. Compared with the 5 V NMOS or PMOS option, the three 1.5 V stacked transistor option has a factor of 1.7 or 2.8 smaller area at the total energy E_{loss} minimum of each option (marked with dots). The two 1.5 V stacked transistor option has a factor of 2.1 or 3.4 smaller area compared to the 5 V NMOS or PMOS option. The area of the charge pump is not included in Fig. 4.12b since it strongly depends on the available capacitor option in the process.

The efficiency benefit of stacked low-voltage transistors is also confirmed by measurement results of the fully integrated resonant SC converter [22, 23] that has

Fig. 4.13 Comparison of the converter efficiency η for different implementation options of the power switches (at $V_{in} = 3.9$ V and $V_{out} = 1.8$ V): two stacked 1.5 V switches versus 5 V PMOS / NMOS transistors directly driven with V_{in} and GND

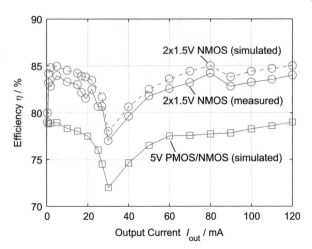

been developed as part of this work (see Chap. 3). Figure 4.13 indicates that an efficiency improvement of up to 6% over the whole current range is achieved with two stacked 1.5 V switches compared to a 5 V PMOS / NMOS implementation.

4.2   Switch Segmentation and Gate Decoupling Circuit

Segmentation of power switches is an established technique that is often used in inductive converters for scaling of the dynamic switching losses at low output power [25–30]. In this work, segmentation of the power switches is used by the switch conductance regulation (SwCR) for controlling the output voltage as described in Sect. 6.1. As a benefit, this also scales the dynamic switching losses which leads to a flat efficiency curve.

However, challenges arise when using stacked low-voltage transistors for the segmented power switches. During the switch conductance regulation, losses at the stacked 2×1.5 V switches do not scale down linearly with the number of activated switches, because of the voltage swing at the parasitic capacitances of the deactivated segments, depicted in Fig. 4.14a. This work proposes two additional switches S_G and S_{GS}, which decouple the gate of the upper transistor and short-circuit the capacitance C_{GS2}, as shown in Fig. 4.14b. When a switch segment is inactive, S_G is off and S_{GS} is on.

Switch S_G is implemented with a 1.5 V PMOS, while S_{GS} is implemented as a transmission gate to guarantee reliable turn-on as shown in Fig. 4.14c. The body diode of S_G will protect the gate and, therefore, also the potential V_1 in the middle of the transistor stack against overvoltage. The switches are small and can be controlled with the existing logic levels SwSel$< x >_{HS}$ on the high-side. The drivers I_1 and I_2 for the switches are supplied from the charge pump on the high-side. The other logic gates are in the other voltage domain defined by $V_{SS,HS}$ and $V_{DD,HS}$.

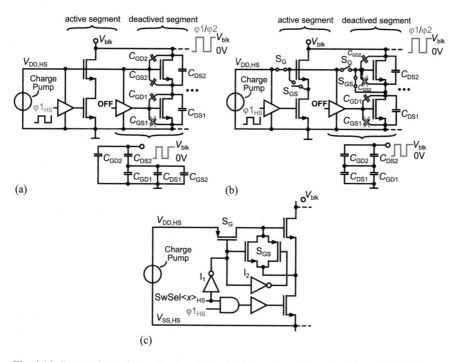

Fig. 4.14 Proposed gate decoupling circuit for effective scaling of the capacitive switching losses: (**a**) without gate decoupling; (**b**) with gate decoupling; (**c**) implementation of gate decoupling switches

The resulting effective capacitance values $C_{eq,w/o}$ and $C_{eq,w/}$ for the turned off segments, without and with decoupling, respectively, can be calculated from the equivalent circuits shown in Fig. 4.14.

$$C_{eq,w/o} = C_{GD} + \frac{C_{DS}(C_{GD} + C_{DS} + C_{GS})}{2C_{DS} + C_{GD} + C_{GS}} \tag{4.12}$$

$$C_{eq,w/} = 2 \cdot \frac{C_{GD} \cdot C_{DS}}{C_{GD} + C_{DS}} \tag{4.13}$$

The relationship between the capacitance values for the used transistors in the process, i.e., $C_{DS} \sim 0.53 \cdot C_{GS}$ and $C_{GD} \sim 0.26 \cdot C_{GS}$, results in the ratio $C_{eq,w/o}/C_{eq,w/} = 1.9$. Thus, the proposed gate decoupling circuit leads to 1.9x lower effective capacitance and significantly lower charge redistribution losses. The effect on the overall efficiency of the proposed converter (Chap. 3) is shown in Fig. 4.15. At low output currents, the gate decoupling circuit leads to significant efficiency improvement compared to the stacked 2×1.5 V switches without gate decoupling, such as 4% at 30 mA.

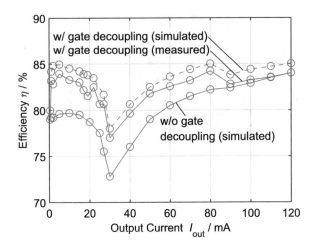

Fig. 4.15 Converter efficiency η with and without the proposed gate decoupling circuit versus output current I_{out} (at $V_{in} = 3.9$ V and $V_{out} = 1.8$ V)

4.3   Level Shifters

Several level shifters are required to shift the necessary control signals for the power switches of the resonant SC converter (ReSC). Different control signal types have to be transferred to the flying high-side domain (Fig. 4.16). Static signals like the 8-bit switch conductance information Sw<0:7> and the back-to-back control signal (see Sect. 4.5.1) and dynamic signals like the clock signals $\varphi 1$ and $\varphi 2$. For this task, different types of level shifters are used: static level shifters and dynamic level shifters. The distinction between static and dynamic level shifter is made based on their duty for the converter operation.

Static level shifters are used to shift control signals with a changing rate far below the switching frequency f_{sw}. For example, the 8-bit switch conductance information Sw<0:7> changes only for regulation purposes. As soon as a steady state is reached, there is no further change of these bits. The level shifters have to hold their output value during any switching transition of the power stage. It has to be robust against fast high-side transitions, which can reach values up to 35 V/ns. For a good efficiency of the whole converter, the power consumption of these level shifters has to be as low as possible during idle state. This is defined as static power consumption for the present application.

For the switch conductance regulation with the segmented power switches, a short propagation delay of the level shifter is important in order to not affect the stability and dynamic response of the control loop.

The dynamic level shifter is used to shift the clock signal φ, which changes for each phase. The dynamic power consumption, required for signal shifting, is important as well as the propagation delay. The shifted clock signals $\varphi 1$ and $\varphi 2$ have to be synchronous for all high-side and low-side switches. A variance of the propagation delay of the level shifter has to be compensated by a dead time to avoid a short between the input voltage and ground. For the proposed resonant SC

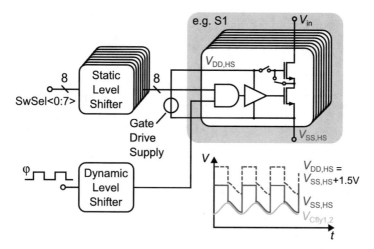

Fig. 4.16 Overview of different level shifter options required for control of the power switches

converter, Fig. 4.16 shows an overview of the different level shifter options together with voltage waveforms during operation.

Several floating level shifters have been published [31–35]. Often, these level shifters are robust and can handle high voltages but at the cost of high power consumption and increased propagation delay. A very popular approach, also in low-power designs, is the use of a cross-coupled level shifter [31, 36]. However, high cross-currents lead to large power consumption and also to high propagation delay. Pulsed level shifters, which use small pulses in combination with a latch at the flying high-side potential $V_{SS,HS}$, can overcome these limitations and lead to a fast and efficient signal transmission. The following sections present two different types of pulsed level shifter topologies. A pulsed cascode level shifter, introduced in Sect. 4.3.1, is used as a static level shifter. For the dynamic level shifters, a capacitive type is used, introduced in Sect. 4.3.2.

4.3.1 Pulsed Cascode Level Shifter

The basic idea is the use of small pulses in combination with a latch at the flying high-side potential $V_{SS,HS}$. Set and reset pulses can be generated on the low-side and efficiently transmitted to the high-side where the latch toggles and afterward keeps its state. A high robustness is achieved due to the double signal paths which enables differential control. There are different ways to design a level shifter using a latch on the flying high-side. A simple pulsed resistive level shifter is presented in [37, 38]. But the used high ohmic resistors are area consuming, and a trade-off with the maximum switching frequency has to be made. An advanced example is presented by [39] and pictured in Fig. 4.17. It is also used in the proposed ReSC

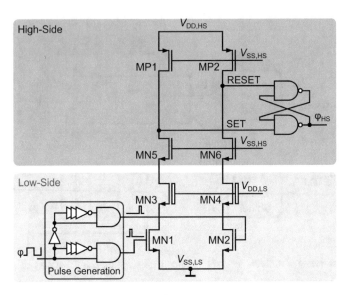

Fig. 4.17 Schematic of the pulsed cascode level shifter

converter. An RS-latch with inverted inputs is formed by two cross-coupled NAND
gates on the high-side. The input signal SET or RESET has to be pulled to a logic
low to set the output to the corresponding value. The two PMOS transistors MP1
and MP2 in Fig. 4.17 operate as pull-up for the inputs SET and RESET of the
NAND latch. With $V_{SS,HS}$ attached to the gates, these transistors are always turned
on and pull the latch input nodes to the high potential $V_{DD,HS}$. The level shifter
would work with any pull-up net. For example, the transistors could be replaced by
resistors, as presented in [37]. But this would lead to a higher area consumption.
The NMOS transistors MN1 and MN2 are used to pull down the SET or RESET
input of the NAND latch. The most simple approach would be the use of only
one 5-V-thick gate transistor for each signal. But the implementation presented
in Fig. 4.17 has several benefits regarding shifting speed and energy. MN1 and MN2
are small 1.5-V-thin gate NMOS transistors. They are faster and have a smaller gate
capacitance than a comparable thick gate transistor with a similar on-resistance.
Transistors MN3 and MN4 are thick gate cascode transistors to protect the low-
voltage input transistors from destructively high drain voltages. These transistors
are the namesake of the circuit. MN5 and MN6 are low-voltage NMOS transistors
on the high-side in common-gate configuration. They amplify the change at the
drain of the high-voltage transistors and enhance the speed of the level shifter [39].
With this, not much charge will be required to change their drain voltages. This
reduces the transition energy, and the propagation delay is not determined by the
drain voltage rise time [39]. To avoid static power consumption, the level shifter
requires short pulses at the input transistors, which are generated with the "Pulse
Generation" block shown at the bottom of the schematic. It should be noted that

the voltage stress on the low-voltage transistors in the high-side voltage domain is a diode forward voltage higher than the supply voltage $V_{DD,HS} - V_{SS,HS}$ due to the forward bias of the bulk-source diodes of MN5 and MN6 during transitions. However, this is acceptable in this application.

4.3.2 Capacitive Level Shifter

The basic idea of the capacitive level shifter is the same as presented for the pulsed cascode level shifter in Sect. 4.3.1, but the implementation of the pulse transmission and the latch is different. The latch of the capacitive level shifter is built with two inverters in positive feedback. The pulse generation and transmission are performed by coupling capacitances between the low-side and the high-side. They act as high-pass and DC decoupling to transmit pulses from the low-side to the high-side whenever there is an edge at the low-side input. One implementation of the capacitive level shifter with a constant high-side potential is proposed in [40]. Adapted for application in the proposed converter, the level shifter is shown in Fig. 4.18. The coupling capacitances C_1 and C_2 transfer edges at the low-side input as galvanic decoupled pulses to the high-side. The inverters I1 and I2 are used to drive the clock input signal φ with a sharp edge for the capacitances to get a strong impulse from the low-side to the high-side. The inverters I3 and I4 form the latch on the high-side. I5 is used to provide a well-driven output signal on the high-side, which is not loaded by the coupling capacitances C_1 and C_2. For the design goal of a low shifting energy, the capacitances C_1 and C_2 are chosen to be as small as possible (both 30 fF) in the given process technology. The design procedure for the inverters I3 and I4 is described in [41]. The output of I3 fights against the pulse transmitted via C_2, and I4 fights against the pulse via C_1. A higher impedance of the inverter

Fig. 4.18 Schematic of the capacitive level shifter

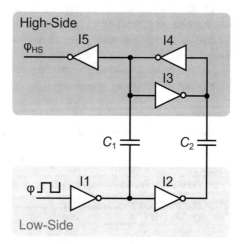

leads to a higher sensitivity of the latch to a pulse transmitted from the low-side
to the high-side but also to a higher noise sensitivity. According to the high-side
supply voltage waveform of $V_{SS,HS}$ and $V_{DD,HS}$ in Fig. 4.16, the high-side potential
changes during the converter operation. As the capacitances C_1 and C_2 have to be
recharged, the inverters have to be strong enough to recharge the capacitances when
a change in the high-side potential occurs. On the other hand, the latch must not
trigger in case of a changing supply voltage. Each change and recharging of the
capacitances couple back to the low-side. The inverters I3 and I4 have to be able to
deliver these recharge currents, and the inverters I1 and I2 have to be strong enough
to keep a constant output while the high-side changes. While reliable operation can
be achieved for the capacitive level shifter in the present application, the flying high-
side and the subsequent recharging of the capacitances lead to losses, even if no
signal change is shifted. This is a disadvantage compared to the pulsed cascode
level shifter.

4.3.3 Comparison of Level Shifter Topologies

Table 4.2 summarizes the key parameters of the presented level shifters. They
were simulated for a clock frequency of 35.5 MHz. In addition, the simulated key
parameters for cross-coupled level shifter [31] are also included for reference. For
the voltage values of the waveform of $V_{SS,HS}$ (Fig. 4.16), the conditions at switch S1
are used were $V_{SS,HS}$ changes between 3.8 V and 1.8 V. The capacitive level shifter
has a very short propagation delay, which is about one order of magnitude shorter
than the pulsed cascode level shifter. It is also the most efficient solution regarding
the shifting energy. Due to the coupling capacitances between the flying high-side
and the static low-side, it needs more power to hold its output against the flying
high-side than the pulsed cascode. The pulsed cascode level shifter represents a good
trade-off between low static power consumption and propagation delay. Since there
are only small parasitic capacitances, it does not need much power for maintaining
its output at a certain state. The power consumption is about one order of magnitude
better than for the capacitive level shifter and within the same order of magnitude as
the cross-coupled level shifter. The cross-coupled level shifter is by far the slowest
and needs much more energy than the other two level shifters for signal transfer.

Table 4.2 Comparison of the presented level shifter topologies

Parameter	Capacitive	Pulsed cascode	Cross-coupled [31]
Static power consumption	800 nW	95 nW	36 nW
Average shifting energy	0.1 pJ	0.8 pJ	2.2 pJ
Propagation delay rise	0.1 ns	0.96 ns	5.4 ns
Propagation delay fall	86 ps	1.07 ns	5.3 ns
Area consumption	255 μm^2	700 μm^2	600 μm^2

Based on these results, the capacitive level shifter is employed as the dynamic level shifter in Fig. 4.16. For the static level shifters, the pulsed cascode level shifter is chosen since it has the lowest static power consumption. Especially regarding the fact that there 58 static level shifters needed, the pulsed cascode level shifter is the most efficient solution for the proposed ReSC converter.

4.4 Gate Drive Supply Generation

In Sect. 4.1, the implementation of the power switches for the ReSC voltage converter is discussed. For the final choice of stacked and segmented 1.5 V NMOS transistors as power switches, a flying supply voltage is needed for each of the seven high-side switches, as shown in Fig. 4.16. This voltage is primarily needed to turn the transistors on and off but also to provide the logic levels for the control signals transmitted from the low-side by the level shifters. This is especially challenging since the reference voltage for each switch is at a different value and changes in a different way during the operation of the ReSC converter. For the overall goal of a highly efficient operation, the generation of the flying supply voltage should be as efficient as possible to further take advantage of the low-voltage transistor stacking over the high-voltage PMOS/NMOS switch directly driven with the input voltage (see Sect. 4.1.3).

A common approach to generate a supply voltage for clocked high-side switches is to use a bootstrap circuit. The basic bootstrapping requires the source of the used high-side NMOS power switch to be at the global reference potential GND during one phase of its clocking. A capacitor with its bottom plate connected to the source of the high-side transistor is then charged with the common low-side supply voltage. This circuit is often used in half-bridges. However, in the proposed as well as in other elaborate power stages [20–22, 38, 42, 43], it cannot be used since the source potentials of the power switches are never at GND during one phase of its clocking.

Floating charge pumps, suitable for arbitrary voltage levels, are introduced in [38] and [44]. They are able to generate a floating high-side potential but at the cost of three diode drops [44] or two diode drops [38]. This is only acceptable when using high turn-on voltages, e.g., in the range of $V_{DD} = 5\,V$. In low-voltage transistor stacking applications, the turn-on voltage has to be derived by a common low-side supply voltage, e.g., $V_{DD} = 1.5\,V$. The voltage drop of a single diode is not tolerable and these solutions cannot be used.

Power stages with a fixed and regular structure may use nested bootstrapping [42, 45] to generate a flying high-side supply or directly use internal voltage rails [18]. Both options cannot be used in the proposed resonant multi-ratio power stage since the voltage levels over each switch vary over the conversion ratios. Due to resonant operation, there are also no fixed voltage rails available during the switching phases $\varphi 1$ and $\varphi 2$. Additionally, nested bootstrap circuits rely on several diodes whose voltage drop is not acceptable with low-voltage transistors.

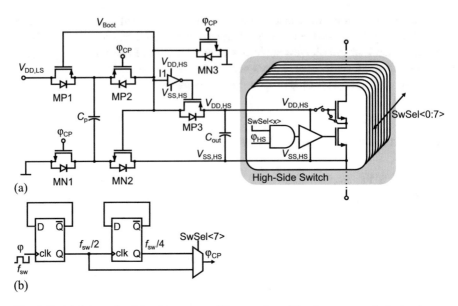

Fig. 4.19 (a) Schematic of the charge pump; (b) generation of the charge pump clock signal φ_{CP}

In order to have a general and efficient solution for the proposed power stages with an input voltage range of $V_{in} = 3\,V - 4.5\,V$, a charge pump circuit [22] has been developed as part of this work based on the bootstrapped switch in [46, 47]. In [46, 47], only one capacitor is used as bootstrap capacitor, which is recharged while the switch is turned off. The switch itself is turned off with GND at the gate. This is not possible for the low-voltage transistors used as power switches for the proposed ReSC converter. Therefore, a second capacitor is introduced to store the charge on the flying high-side and provide a stable supply voltage to turn the switches on and off. The charge pump circuit as an extended version of the bootstrapped switch circuit presented in [46] is shown in Fig. 4.19.

When the low-side clock signal φ_{CP} is high, MN1 and MN3 are turned on. The bottom plate of the pump capacitor C_p is connected to GND with MN1, while MN3 pulls the node voltage V_{boot} and thereby the gate of the PMOS transistor MP1 to GND. MP1 is turned on and C_p is charged from the low-side supply $V_{DD,LS}$. The other transistors MN2, MP2, and MP3 are turned off. When φ_{CP} becomes low, MN3 and MN1 are turned off, while MP2 is turned on. In the first step, it connects V_{boot} to the top plate of the pump capacitor C_p where the potential is $V_{DD,LS}$. With that, MP1 is turned off. The pump capacitor C_p is now completely disconnected from the low-side supply. At the same time, MN2 is turned on. C_p is connected between gate and source of MN2 and keeps it turned on while MN2 pushes the bottom plate of C_p to the high-side reference voltage $V_{SS,HS}$. Neglecting losses, the top plate of C_p and thereby V_{boot} is now at $V_{SS,HS} + V_{DD,LS}$. This is exactly the desired value for the high-side supply $V_{DD,HS}$. If the high-side supply is already present, the inverter I1 turns transistor MP3 on, and the capacitor C_{out} is recharged by C_p. During the start-up of the charge pump circuit, the body diode of MP3 provides the charge flow from C_p to C_{out}. When φ_{CP} rises again, MP2 is turned off and disconnects the two capacitors, MN3 pulls V_{boot} down to GND, and I1 turns MP3 off. The high-side is disconnected and the pump capacitor C_p is recharged at the low-side. One advantage of this concept lies in the self-timing since there is only one GND referred clock signal required to control the charge pump. All other transistors are turned on or off at the right time with the node voltage V_{boot}. For the implementation of this charge pump, all transistors have to be 5 V transistors to tolerate the resulting gate-source and drain-source voltages. When φ is high, the source of the PMOS transistor of I1 is at $V_{DD,HS}$, and the gate is pulled to GND with MN3. The application for this charge pump is thereby limited to $V_{DD,HS,max} \leq V_{GS,max,PMOS}$. Within this limit, the proposed charge pump is a general solution which works for all possible voltage waveforms.

For the sizing of the integrated capacitors, a trade-off between efficiency, area consumption, output voltage ripple, and output voltage drop has to be made. The pump capacitor C_p is set to be 75 pF, which provides enough charge flow to provide a high-side supply voltage always greater than 1.3 V. For low area consumption, the output capacitor C_{out} is also set to 75 pF which leads to a maximum output voltage ripple of 150 mV. Both capacitors are implemented as MOS capacitors.

Depending on how many power switch segments are active, the load of the charge pump varies. In order to scale down the frequency-dependent losses, mainly the capacitive bottom-plate losses, the switch conductance information Sw<0:7> is used to scale the charge pump frequency φ_{CP} as shown in Fig. 4.19b. At higher loads half of the converter switching frequency $\varphi_{CP} = f_{sw}/2$ is used. At lower loads (SwSel<7>=0), φ_{CP} runs at $f_{sw}/4$.

Figure 4.20a shows the simulated efficiency η_{CP} of the proposed charge pump versus 8-bit switch conductance information SwSel<0:7>, which corresponds to the output power. The scaling of the charge pump frequency φ_{CP} leads to a significant improvement of the efficiency for smaller loads (SwSel<0:7> $\leq$ 128). It also leads to a flatter course of the charge pump output voltage $V_{DD,HS} - V_{SS,HS}$, shown in Fig. 4.20b. The resulting switch resistance R_{sw} is shown in Fig. 4.20c. At lower loads, the resistance R_{sw} of a charge pump-supplied switch is lower compared to a fixed turn-on voltage of 1.3V. This is due to the fact that the generated output voltage $V_{DD,HS} - V_{SS,HS}$ by the charge pump increases at lower loads.

The final layout of the charge pump as part of a high-side switch is shown in Fig. 4.21. This full implementation requires an area of approximately 64,000 µm, including the stacked and segmented transistors with the corresponding driver stages, all level shifters, and the charge pump for the high-side supply voltage. The transistors used as power switch cover less than 18 % of the area of the whole block. The eight static level shifters require an area of about 11,500 µm, around the same as used for the total power switch itself. The charge pump is the largest part by far. It consumes slightly less than half of the area due to the large capacitors and the 5 V transistors.

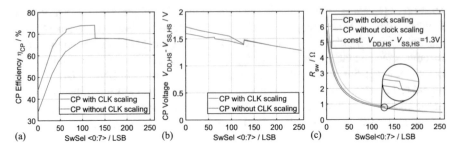

Fig. 4.20 (**a**) Comparison of charge pump efficiency η_{CP}; (**b**) output voltage $V_{DD,HS} - V_{SS,HS}$; (**c**) switch resistance R_{sw} with and without charge pump frequency scaling

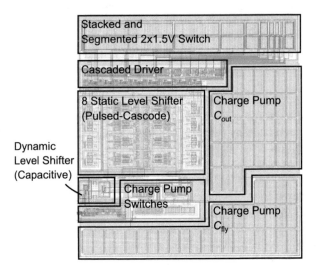

Fig. 4.21 Layout of the charge pump as part of a high-side switch

4.5 Power Stage Implementation

4.5.1 Power Switches

Depending on the location in the power stage, different options for the power switches are implemented. In order to discuss these, the different conversion ratios of the power stage are shown in Fig. 4.22 together with the respective configuration of the power switches. The figure is extended by the voltage waveform of the switching node voltage V_{sw}. In average, V_{sw} has the same value as the output voltage V_{out} but with a half-cosine-shaped ripple. The amplitude of the ripple can reach several 100 mV depending on the present input/output voltage and output current conditions. Simulations show that the power switches have to block voltages up to 3.5 V due to the high resonant voltage swings at the flying capacitors during

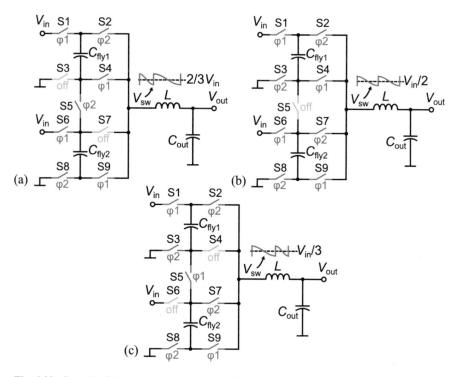

Fig. 4.22 Control of the power switches depending on the conversion ratio N: **(a)** $N=2/3$; **(b)** $N=1/2$; **(c)** $N=1/3$

operation. This is handled by stacking of two 1.5 V low-voltage transistors as discussed in Sect. 4.1.

Figure 4.23a shows the implementation of the two low-side switches S3 and S8. They are supplied with the low-side supply voltage $V_{DD,LS}$ and no charge pump is required. There is no need for level shifters, and the switches can be directly controlled with the low-side control signals. The switch logic block controls the power switch with the correct control signals ($\varphi 1_{sw}/\varphi 2_{sw}/$off) depending on the ratio information N. A short delay is inserted with the delay chain I1 which compensates the propagation delay of the dynamic level shifters of the other high-side transistors. This ensures that the clock signals $\varphi 1_{sw}$ and $\varphi 2_{sw}$ are synchronous for all power switches in the power stage.

The implementation of a standard high-side switch (S1, S2, S5, S6, S9) is shown in Fig. 4.23b. The different control signals have to be transferred to the flying high-side domain via level shifters (see Sect. 4.3). Eight static level shifters are used to shift the 8-bit switch conductance information Sw<0:7>, while one dynamic level shifter is used to shift the clock signals $\varphi 1_{sw}$ and $\varphi 2_{sw}$ or the information whether the power switch is off.

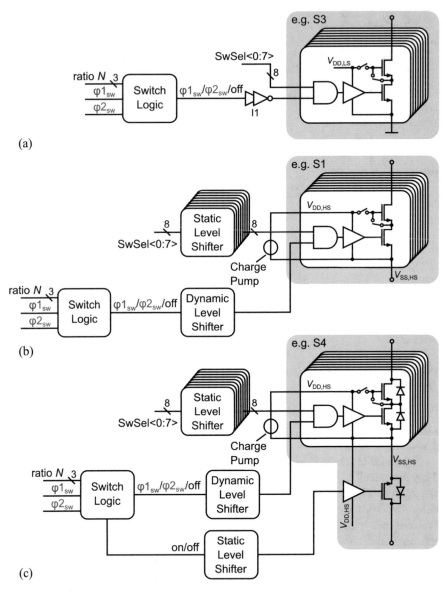

Fig. 4.23 Implementation of segmented power switches with two stacked 1.5 V NMOS transistors and necessary driver circuits: (**a**) implementation of low-side power switch; (**b**) implementation of high-side power switch; (**c**) implementation of high-side power switch with additional back-to-back switch

The reconfigurability of the ReSC converter stage can result in different voltage polarities across the power switches. Since bulk and source of the NMOS transistors are connected, only positive drain-source voltages V_{DS} can be blocked. If the

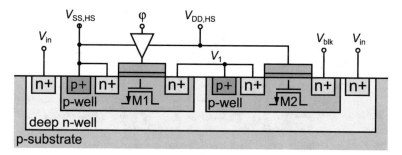

Fig. 4.24 Cross section of the layout of the implemented two stacked 1.5 V NMOS transistors

source and bulk rise above the drain potential, the intrinsic body diode can become conductive. Considering this, the switches S4 and S7 are at risk. In ratio 1/3 (Fig. 4.22a) when S4 is always turned off, it has to block a positive V_{DS} in phase $\varphi 1$ and a negative V_{DS} in phase $\varphi 2$. The same is true for switch S7 in ratio 2/3 (Fig. 4.22c). These two switches need to be implemented in a back-to-back configuration in order to prevent a conductive path through a body diode. The back-to-back configuration is only needed when the ReSC converter is operated within a conversion ratio where the corresponding switch is always turned off. In that case, the back-to-back transistor is only switched when the conversion ratio changes and is not clocked with the operation frequency of the converter. It can be very large to provide a low additional $R_{DS,on}$ for the switch implementation, since the gate charge losses of this transistor are negligible. Figure 4.23c shows the implementation of high-side power switch with an additional back-to-back transistor for the switches S4 and S7. In the ratio 1/3, the additional back-to-back transistor of switch S4 is turned off, while in ratio 2/3 the additional back-to-back transistor of switch S7 is turned off. Since these control signals are set at a rate, which is far below the switching frequency, static level shifters are used in order to reduce the static power consumption (see Table 4.2). In the blocking state, the two body diodes of the stacked switches reduce the blocking voltage by approximately 1.5 V. Thus, it is sufficient to use only a single transistor as an additional back-to-back switch.

Figure 4.24 shows the cross section of the implemented stacked power switches. The naming conventions are taken from Fig. 4.3. To limit the maximum gate-bulk voltages, the source and bulk of every transistor are shorted. Therefore, the p-wells have to be separated from each other. The transistors are surrounded and isolated by a deep n-well biased at the highest potential (V_{in}) which presents capacitive coupling into the substrate.

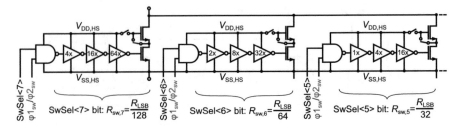

Fig. 4.25 Implementation of gate driver for the segmented power switches

4.5.2   Gate Driver

The gate driver is responsible for a fast charging and discharging of the gate capacitances of the power switches. To avoid considerable transition losses in the power switches, the gate driver strength is designed such that the turn-on transition is at least 80 times faster than the clock phase $\varphi 1$ or $\varphi 2$. Therefore, the gate driver has to be able to charge and discharge the gate capacitance within 180 ps. During fast transitions of the blocking voltage, the gate driver also has to be strong enough to sink the coupling currents through the drain-gate capacitances of the NMOS transistors while keeping the transistor in its switching state. A short propagation delay is achieved by a tapered design of fast switching 1.5 V standard inverter stages with different driver strengths [48–50]. A taper factor of $\beta = 4$ is used, which leads to a propagation delay of 250 ps for the gate driver. Together with the delay of the dynamic level shifter, this results in short total propagation delay of 350 ps. A more complex implementation with separate branches to avoid cross-conduction in the last driver stage is not necessary. The threshold voltage V_{th} of the used NMOS and PMOS transistors is about 0.6 V, which results in a very short time slot of a few picoseconds when both transistors of an inverter are conducting. Thus, the additional losses of the resulting cross-conduction are negligible. Each of the segmented power switches has its own gate driver according to Fig. 4.25. The driver strengths scale directly with the size of the switches in each segment. This ensures the same propagation delay for each segment, which leads to a synchronous switching behavior of all transistor segments that form the power switch.

References

1. Neveu, F., et al.: A 100 MHz 91.5% Peak efficiency integrated buck converter with a Three-MOSFET cascode bridge. IEEE Trans. Power Electron. **31**(6), 3985–3988 (2016). ISSN: 1941-0107. https://doi.org/10.1109/TPEL.2015.2502058
2. Bezerra, P.A.M., et al.: Analysis and comparative evaluation of stacked-transistor half-bridge topologies implemented with 14 nm bulk CMOS technology. In: 2017 IEEE 18th Workshop on Control and Modeling for Power Electronics (COMPEL), pp. 1–8 (2017). https://doi.org/10.1109/COMPEL.2017.8013307

3. Kursun, V., et al.: High input voltage step-down DCDC converters for integration in a low voltage CMOS process. In: International Symposium on Signals, Circuits and Systems. Proceedings, SCS 2003. (Cat. No.03EX720), pp. 517–521 (2004). https://doi.org/10.1109/ISQED.2004.1283725

4. Kursun, V., et al.: Cascode buffer for monolithic voltage conversion operating at high input supply voltages. In: 2005 IEEE International Symposium on Circuits and Systems, Vol. 1, pp. 464–467 (2005). https://doi.org/10.1109/ISCAS20051464625

5. Burton, E.A., et al.: FIVR – fully integrated voltage regulators on 4th generation intel® core™ SoCs. In: 2014 IEEE Applied Power Electronics Conference and Exposition – APEC 2014, pp. 432–439 (2014). https://doi.org/10.1109/APEC2014.6803344

6. Krishnamurthy, H.K., et al.: A digitally controlled fully integrated voltage regulator with on-die solenoid inductor with planar magnetic core in 14-nm tri-gate CMOS. IEEE J. Solid State Circuits 53(1), 8–19 (2018). ISSN: 0018-9200. https://doi.org/10.1109/JSSC.2017.2759117

7. Greenhill, D., et al.: A 330 MHz 4-way superscalar microprocessor. In: 1997 IEEE International Solids-State Circuits Conference. Digest of Technical Papers, pp. 166–167 (1997). https://doi.org/10.1109/ISSCC.1997.585318

8. Serneels, B., Steyaert, M., Dehaene, W.: A 237mW aDSL2+ CO line driver in standard 1.2 V 0.13μm CMOS. In: 2007 IEEE International Solid-State Circuits Conference. Digest of Technical Papers, pp. 524–619 (2007). https://doi.org/10.1109/ISSCC.2007.373525

9. Serneels, B., Steyaert, M., Dehaene, W.: A 5.5 V SOPA line driver in a standard 1.2 V 0.13 μm CMOS technology. In: Proceedings of the 31st European Solid-State Circuits Conference, 2005. ESSCIRC 2005, pp. 303–306 (2005). https://doi.org/101109/ESSCIR20051541620

10. Serneels, B., et al.: A high-voltage output driver in a 2.5V 0.25μm CMOS technology. IEEE J. Solid State Circuits 40(3), 576–583 (2005). ISSN: 1558-173X. https://doi.org/10.1109/JSSC.2005.843599

11. Singh, G.: A high speed 3.3V IO buffer with 1.9V tolerant CMOS process. In: Proceedings of the 24th European Solid-State Circuits Conference, 128–131 (1998). https://doi.org/10.1109/ESSCIR.1998.186225

12. Bradburn, S.R., Hess, H.L.: An integrated high-voltage buck converter realized with a low-voltage CMOS process. In: 2010 53rd IEEE International Midwest Symposium on Circuits and Systems, pp. 1021–1024 (2010). https://doi.org/10.1109/MWSCAS.2010.5548815

13. Annema, A., Geelen, G.J.G.M., de Jong, P.C.: 5.5V I/O in a 2.5V 0.25μm CMOS Technology. IEEE J. Solid State Circuits 36(3), 528–538 (2001). ISSN: 1558-173X. https://doi.org/10.1109/4.910493

14. Oestman, K.B., Jaervenhaara, J.K.: A rapid switch bridge selection method for fully integrated DCDC buck converters. IEEE Trans. Power Electron. 30(8), 4048–4051 (2015). ISSN: 1941-0107. https://doi.org/10.1109/TPEL.2014.2384915

15. Renz, P., Kaufmann, M., Lueders, M., Wicht, B.: Switch stacking in power management ICs. IEEE J. Emerg. Sel. Top. Power Electron. (in press 2020). https://doi.org/10.1109/JESTPE.2020.3012813

16. Hardy, C., Le, H.: A 10.9W 93.4%-efficient (27W 97%-efficient) flying-inductor hybrid DCDC converter suitable for 1-cell (2-Cell) battery charging applications. In: 2019 IEEE International Solid- State Circuits Conference – (ISSCC), pp. 150–152 (2019). https://doi.org/10.1109/ISSCC.2019.8662432

17. Kim, W., Brooks, D., Wei, G.: A fully-integrated 3-level DCDC converter for nanosecond-scale DVFS. IEEE J. Solid State Circuits 47(1), 206–219 (2012). ISSN: 0018-9200. https://doi.org/10.1109/JSSC.2011.2169309

18. Liu, W., et al.: A 94.2%-peak-efficiency 1.53A direct-battery-hook- up hybrid dickson switched-capacitor DCDC converter with wide continuous conversion ratio in 65nm CMOS. In: 2017 IEEE International Solid- State Circuits Conference (ISSCC), pp. 182–183 (2017). https://doi.org/10.1109/ISSCC20177870321

19. Schaef, C., Stauth, J.T.: A 3-phase resonant switched capacitor converter delivering 7.7W at 85% efficiency using 1.1nH PCB trace inductors. IEEE J. Solid State Circuits 50(12), 2861–2869 (2015). ISSN: 1558-173X. https://doi.org/10.1109/JSSC.2015.2462351

20. Schaef, C., Din, E., Stauth, J.T.: A digitally controlled 94.8%- peak-efficiency hybrid switched-capacitor converter for bidirectional balancing and impedance-based diagnostics of lithium-ion battery arrays. In: 2017 IEEE International Solid-State Circuits Conference (ISSCC), pp. 180–181 (2017). https://doi.org/10.1109/ISSCC.2017.7870320

21. Lutz, D., Renz, P., Wicht, B.: A 10mW fully integrated 2- to-13V-input buck-boost SC converter with 81.5% peak efficiency. In: 2016 IEEE International Solid-State Circuits Conference (ISSCC), pp. 224–225 (2016). https://doi.org/10.1109/ISSCC.2016.7417988

22. Renz, P., et al.: A fully integrated 85%-peak-efficiency hybrid multi ratio resonant DCDC converter with 3.0-to-4.5V input and $500\mu A$ -to- 120mA load range. In: 2019 IEEE International Solid- State Circuits Conference – (ISSCC), pp. 156–158 (2019). https://doi.org/10.1109/ISSCC.2019.8662491

23. Renz, P., et al.: A 3-ratio 85% efficient resonant SC converter with on-chip coil for Li-Ion battery operation. IEEE Solid-State Circuits Lett. 2(11), 236–239 (2019). ISSN: 2573-9603. https://doi.org/10.1109/LSSC.2019.2927131

24. Bandyopadhyay, S., Ramadass, Y.K., Chandrakasan, A.P.: 20 μA to 100mA DC–DC converter with 2.8–4.2V battery supply for portable applications in 45nm CMOS. IEEE J. Solid State Circuits 46(12), 2807–2820 (2011). ISSN: 1558-173X. https://doi.org/10.1109/JSSC.2011.2162914

25. Park, S., et al.: A PWM buck converter with load-adaptive power transistor scaling scheme using analog-digital hybrid control for high energy efficiency in implantable biomedical systems. IEEE Trans. Biomed. Circuits Syst. 9(6), 885–895 (2015). ISSN: 1940-9990. https://doi.org/10.1109/TBCAS.2015.2501304

26. Trescases, O., Wen, Y.: A survey of light-load efficiency im- provement techniques for low-power DCDC converters. In: 8th International Conference on Power Electronics ECCE Asia, pp. 326–333 (2011). https://doi.org/10.1109/ICPE.2011.5944617

27. Michal, V.: Peak-efficiency detection and peak-efficiency tracking algorithm for switched-mode DC–DC power converters. IEEE Trans. Power Electron. 29(12), 6555–6568 (2014). ISSN: 1941-0107. https://doi.org/10.1109/TPEL.2014.2304491

28. Luo, P., et al.: Digital assistant current sensor for PWM DCDC converter with segmented output stage. In: 2013 International Conference on Communications, Circuits and Systems (ICCCAS), vol. 2, pp. 358–361 (2013). https://doi.org/10.1109/ICCCAS.2013.6765356

29. Jain, R., et al.: Conductance modulation techniques in switched- capacitor DCDC converter for maximum-efficiency tracking and ripple mitigation in 22nm tri-gate CMOS. In: Proceedings of the IEEE 2014 Custom Integrated Circuits Conference, pp. 1–4 (2014). https://doi.org/10.1109/CICC.2014.6946052

30. Wu, C., et al.: Digital buck converter with switching loss reduction scheme for light load efficiency enhancement. IEEE Trans. Very Large Scale Integr. VLSI Syst. 25(2), 783–787 (2017). ISSN: 1557-9999. https://doi.org/10.1109/TVLSI.2016.2592537

31. Moghe, Y., Lehmann, T., Piessens, T.: Nanosecond delay floating high voltage level shifters in a 0.35 μm HV-CMOS technology. IEEE J. Solid State Circuits 46(2), 485–497 (2011). ISSN: 0018-9200. https://doi.org/10.1109/JSSC.2010.2091322

32. Wittmann, J., Rosahl, T., Wicht, B.: A 50V high-speed level shifter with high dv/dt immunity for multi-MHz DCDC converters. In: ESS- CIRC 2014 – 40th European Solid State Circuits Conference (ESSCIRC), pp. 151–154 (2014). https://doi.org/10.1109/ESSCIRC.2014.6942044

33. Liu, D., Hollis, S.J., Stark, B.H.: A new design technique for sub-nanosecond delay and 200V/ns power supply slew-tolerant floating voltage level shifters for GaN SMPS. IEEE Trans. Circuits Syst. I Regul. Pap. 66(3), 1280–1290 (2019). ISSN: 1549-8328. https://doi.org/10.1109/TCSI.2018.2878668

34. Larsen, D.O., et al.: High-voltage pulse-triggered sr latch level- shifter design considerations. In: 2014 NORCHIP, pp. 1–6 (2014). https://doi.org/10.1109/NORCHIP.2014.7004737

35. Song, M.K., et al.: A 20V 8.4W 20MHz four-phase GaN DCDC converter with fully on-chip dual-SR bootstrapped GaN FET driver achieving 4ns constant propagation delay and 1ns switching rise time. In: 2015 IEEE International Solid-State Circuits Conference – (ISSCC) Digest of Technical Papers, pp. 1–3 (2015). https://doi.org/10.1109/ISSCC.2015.7063046

36. Koo, K-H., et al.: A new level-up shifter for high speed and wide range interface in ultra deep sub-micron. In: 2005 IEEE International Symposium on Circuits and Systems, vol. 2, pp. 1063–1065 (2005). https://doi.org/10.1109/ISCAS2005.1464775

37. Lutz, D., Renz, P., Wicht, B.: A 120/230Vrms-to-3.3V micro power supply with a fully integrated 17V SC DCDC converter. In: ESSCIRC Conference 2016: 42nd European Solid-State Circuits Conference, pp. 449–452 (2016). https://doi.org/10.1109/ESSCIRC.2016.7598338

38. Lutz, D., Renz, P., Wicht, B.: An integrated 3-mW 120/230-V AC mains micropower supply. IEEE J. Emerg. Sel. Top. Power Electron. 6(2), 581–591 (2018). ISSN: 2168-6777. https://doi.org/10.1109/JESTPE.2018.2798504

39. Lehmann, T.: Design of fast low-power floating high-voltage level-shifters. Electron. Lett. 50(3), 202–204 (2014). ISSN: 0013-5194. https://doi.org/10.1049/el.2013.2270

40. Tanzawa, T., et al.: High-voltage vransistor scaling circuit tech-niques for high-density negative-gate channel-erasing NOR flash memories. IEEE J. Solid State Circuits 37(10), 1318–1325 (2002). ISSN: 0018-9200. https://doi.org/10.1109/JSSC.2002.803045

41. Zheng, W-M., et al.: Capacitive floating level shifter: modeling and design. In: TENCON 2015 – 2015 IEEE Region 10 Conference, pp. 1–6 (2015). https://doi.org/10.1109/TENCON.2015.7373013

42. Kesarwani, K., Stauth, J.T.: The direct-conversion resonant switched capacitor architecture with merged multiphase interleaving: cost and performance comparison. In: 2015 IEEE Applied Power Electronics Conference and Exposition (APEC), pp. 952–959 (2015). https://doi.org/10.1109/APEC.2015.7104464

43. Schaef, C., Kesarwani, K., Stauth, J.T.: A variable-conversion- ratio 3-phase resonant switched capacitor converter with 85% efficiency at 0.91W/mm^2 using 1.1nH PCB-trace inductors. In: 2016 IEEE International Solid-State Circuits Conference – (ISSCC) Digest of Technical Papers, pp. 1–3 (2015). https://doi.org/10.1109/ISSCC.2015.7063075

44. Park, S., Jahns, T.M.: A self-boost charge pump topology for a gate drive high-side power supply. IEEE Trans. Power Electron. 20(2), 300–307 (2005). ISSN: 0885-8993. https://doi.org/10.1109/TPEL2004843013

45. Dougherty, C.M., et al.: A 10V fully-integrated switched-mode step-up piezo drive stage in 0.13 μm CMOS using nested-bootstrapped switch cells. IEEE J. Solid State Circuits 51(6), 1475–1486 (2016). ISSN: 0018-9200. https://doi.org/10.1109/JSSC.2016.2551221

46. Siragusa, E., Galton, I.: A digitally enhanced 1.8-V 15-Bit 40-MSample/s CMOS pipelined ADC. IEEE J. Solid State Circuits 39(12), 2126–2138 (2004). ISSN: 0018-9200. https://doi.org/10.1109/JSSC.2004.836230

47. Dessouky, M., Kaiser, A.: Input switch configuration suitable for rail-to-rail operation of switched OP amp circuits. Electron. Lett. 35(1), 8–10 (1999). ISSN: 0013-5194. https://doi.org/10.1049/el:19990028

48. Li, N.C., Haviland, G.L., Tuszynski, A.A.: CMOS tapered buffer. IEEE J. Solid-State Circuits 25(4), 1005–1008 (1990). https://doi.org/10.1109/4.58293

49. Wittmann, J., Wicht, B.: MHz-converter design for high conversion ratio. In: 2013 25th International Symposium on Power Semiconductor Devices IC's (ISPSD), pp. 127–130 (2013). https://doi.org/10.1109/ISPSD.2013.6694445

50. Wittmann, J., et al.: An 18V input 10MHz buck converter with 125ps mixed-signal dead time control. IEEE J. Solid State Circuits 51(7), 1705–1715 (2016). ISSN: 1558-173X. https://doi.org/10.1109/JSSC.2016.2550498

Chapter 5
Integrated Passives

This chapter presents concepts, implementation details, and achievable parameters of integrated passives for resonant SC converters.

Section 5.1 covers the design of loss optimized integrated flying capacitors. The modeling and implementation of integrated inductors are addressed in Sect. 5.2. In addition, fully integrated and off-chip inductors are compared.

5.1 Design of Loss Optimized Integrated Capacitors

Capacitors are a key component of resonant SC converters. Flying capacitors C_{fly} provide the charge transfer, while the output capacitor C_{out} is responsible for damping of the output voltage V_{out} ripple of the converter. With the integration of capacitors, the main challenges are the parasitic capacitances, especially the parasitic bottom plate capacitance, and the capacitance density. The capacitor density $C_{\square}$ determines the area consumption A and thus the cost of integrated capacitor and is defined as

$$C_{\square} = \frac{C}{A}. \tag{5.1}$$

The parasitic bottom plate loss is related to the quality factor α, defined in Eq. 2.31. It is the ratio of the parasitic bottom capacitance C_{BP} with respect to the actual useful capacitance C_{fly}. It defines the quality of an capacitor and has a large impact on the converter's efficiency.

The capacitor resistance is also a critical parameter. It introduces additional power loss and an additional voltage ripple at the output.

© The Editor(s) (if applicable) and The Author(s), under exclusive license to
Springer Nature Switzerland AG 2021
P. Renz, B. Wicht, *Integrated Hybrid Resonant DCDC Converters*,
https://doi.org/10.1007/978-3-030-63944-0_5

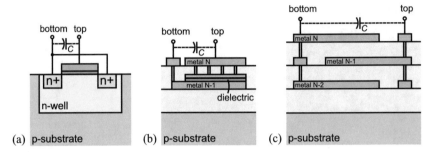

Fig. 5.1 Overview of different capacitor types: **(a)** MOS capacitor; **(b)** MIM capacitor; **(c)** MOM capacitor

In the following, three common capacitor options in a standard CMOS process are analyzed (Fig. 5.1): MOS capacitors, MIM capacitors, and MOM capacitors. Capacitor types like trench capacitors [1–3] or ferroelectric capacitors [4, 5] are not compared here since they are not available in standard CMOS since they require additional expensive post process steps.

The structure of a MOS capacitor is shown in Fig. 5.1a. It utilizes the gate capacitance of a MOS transistor. The top plate of the capacitor is formed by the gate electrode (polycrystalline silicon), while the bottom plate is formed by the underlying n-well. The capacitance value is basically set by the thin gate oxide dielectric. The capacitance of the MOS capacitor depends on the voltage (bias) on the gate.

In accumulation, typically a positive voltage is applied where the positive charge on the gate attracts negative electrons from the substrate to the oxide-semiconductor interface. There, typical values for the capacitance density are in the range of $C_\square = 2 - 8\,\text{fF}/\mu\text{m}^2$ depending on the thickness of the gate oxide (e.g., 1.5-V-thin gate or 5-V-thick gate). Since the capacitor is close to the substrate, higher values for the quality factor (lower quality) of around $\alpha = 5\%$ are obtained.

A MIM capacitor is formed like a traditional plate-type capacitor in the upper layers of the metal stack, depicted in Fig. 5.1b. To increase the capacitance density, a standard metal layer and an additional layer are used. Thereby, a smaller spacing between the plates is introduced. An additional dielectric with a high permittivity leads to a high capacitance density of $C_\square = 1 - 2\,\text{fF}/\mu\text{m}^2$. Due to the location in the upper metal layers, a quality factor of $\alpha = 1 - 2\%$ is obtained.

MOM capacitors are formed with finger-type structures out of several metal layers according to Fig. 5.1c. The capacitance density increases with the number of metal layers but also the capacitive coupling to the substrate, which reduces the quality factor α of the capacitor. Compared to a MIM capacitor, no additional mask is required, and higher voltage ratings are achieved. Typical values for the capacitance density are in the range of $C_\square = 0.2 - 1\,\text{fF}/\mu\text{m}^2$.

The flying capacitors C_{fly1} and C_{fly2} of the resonant SC converter are implemented with different capacitor options for comparison. MOS and MIM capacitors

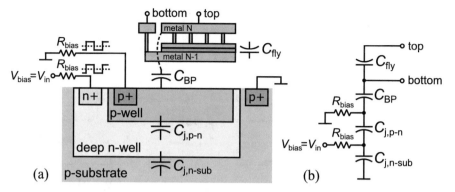

Fig. 5.2 (**a**) Implementation of flying capacitors C_{fly} with MIM capacitors. (**b**) Equivalent circuit

are used. MIM capacitors are preferred over MOM since they have higher capacitance density and lower parasitic coupling at the cost of an additional mask. In the following, implementation details are presented together with a comparison of the different options.

5.1.1 Implementation of MIM Capacitors

The implementation of the MIM capacitor as a flying capacitor C_{fly} is shown in Fig. 5.2a. It is located between the metal layers 5 and 6. It is a common technique for MOS capacitors to reduce the parasitic bottom plate coupling by suitable biasing the wells of the MOS capacitors [6–8]. This leads to additional junction capacitances in series, which minimizes the overall effective bottom plate capacitance. For MIM caps, this technique was first shown in [9–11] but only with one single n-well. In order to further improve the quality factor of the MIM capacitor, an underlying series of a high-resistively biased p-well (biased at GND) and a deep n-well (biased at V_{in}) minimizes the influence of the parasitic bottom plate capacitance [12, 13]. This leads to the equivalent circuit shown in Fig. 5.2b. The series connection of C_{BP}, $C_{j,p-n}$, and $C_{j,n-sub}$ leads to a significant lower effective bottom plate capacitance.

Figure 5.3a shows the simulated junction capacitances $C_{j,p-n}$ and $C_{j,n-sub}$ versus the bias voltage. Larger bias voltages lead to lower capacitance values since the depletion region is increased. In the implemented converter, the junction capacitors are biased with the input voltage of the converter which is in the range of $V_{in} = 3\,V - 4.5\,V$. This technique reduces the quality factor from $\alpha = 2\%$ (without wells) to $\alpha = 1.32 - 1.27\%$, which can be seen in Fig. 5.3b. Higher values of the biasing voltage V_{bias} achieve only a small improvement. In addition, V_{bias} would have to be generated by an additional charge pump circuit, which in turn reduces the efficiency benefit.

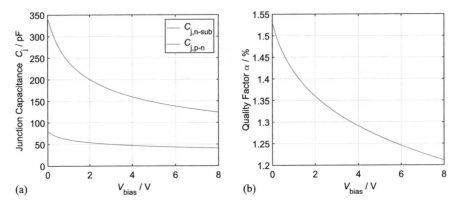

Fig. 5.3 (a) Simulated junction capacitances $C_{j,p-n}$ and $C_{j,n-sub}$ for the p-well and deep n-well versus bias voltage V_{bias}; (b) resulting quality factor of the MIM capacitor

The bias resistor $R_{bias} = 1\,M\Omega$ is implemented with a poly high-sheet resistance and leads to a sufficient high time constant ($\tau \gg 1/f_{sw}$) for effectively decoupling the capacitances. [8] proposes to implement R_{bias} as pseudo resistor of two front-to-front diode-connected PMOS transistors. This technique is not applied in this design since it reduced the effective biasing voltage V_{bias} by roughly one diode drop. The smaller depletion region would lead to higher junction capacitances (see Fig. 5.3a), which counteract the benefit of this biased-well technique compared with the implemented poly resistance. Additionally, there is no area benefit when a high-resistance option is available in the process, as is the case in this work.

For a further improvement of the capacitor quality, the flying capacitors are flipped [6–13] since the parasitic top plate capacitance is slightly smaller than the parasitic bottom plate. Thus, the larger parasitic capacitor contributes to the charge transport to the output (see Sect. 2.3.3).

Charge recycling of the parasitic bottom plate capacitance during the dead time between the phases $\varphi 1$ and $\varphi 2$ is proposed in [14–17]. There, the converters operate at low input voltages $<2\,V$ and low switching frequencies which significantly simplifies the implementation of the required additional switches. In the proposed converter, charge recycling techniques are not reasonable due to the high input voltages and the high switching frequency and the resulting short dead-time. In order to achieve a sufficiently low RC time constant for the short dead-time so that the charge recycling can take place, the switches must be very large. For the proposed implementation, the associated switching losses would completely compensate the efficiency benefit.

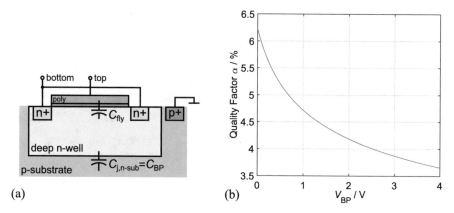

Fig. 5.4 (**a**) Implementation of flying capacitors C_{fly} with MOS capacitors; (**b**) quality factor α versus bottom-plate voltage V_{BP}

5.1.2 Implementation of MOS Capacitors

The implemented accumulation-based MOS capacitor is shown in Fig. 5.4a. For the implementation of the flying capacitors C_{fly1} and C_{fly2}, a 5 V thick-oxide MOS cap with a capacitance density of $C_\square = 2\,\mathrm{fF}/\mu\mathrm{m}^2$ is used. The gate electrode is the top plate of the capacitor, while the bottom plate is formed by the underlying deep n-well. Compared to p-type solutions [6], the low doping concentration of the deep n-well results in a low junction capacitance $C_{j,n-sub}$ for the deep n-well to p-substrate junction, which forms the bottom plate capacitor C_{BP} in this case. The quality factor α of the MOS capacitor depends on the voltage V_{BP} at the bottom plate of the flying capacitor as shown in Fig. 5.4b. If the converter is in 1/2 configuration, the average voltage V_{BP} is in the range $V_{BP} = 0\,\mathrm{V} \ldots V_{in}/2$, which leads to quality factors of $\alpha = 4.2 - 6\%$.

5.1.3 Comparison Between MIM and MOS Capacitor Options

Figure 5.5 shows a part of the implemented test chips with the power stage and the different options for the flying capacitors C_{fly1} and C_{fly2} with both MIM and MOS capacitors. The MIM capacitors occupy an area of $2 \cdot 0.85\,\mathrm{mm}^2$, while MOS capacitors require a smaller area of $2 \cdot 0.44\,\mathrm{mm}^2$ due to the higher capacitance density $C_\square$. A comparison of the measured converter efficiency with the different capacitor options is shown in Fig. 5.6. If the converter operates with switch conductance regulation (SwCR) (see Sect. 3.2.2), the MOS option results in up to 5% lower efficiency at low output currents (Fig. 5.6a). This is due to the fact that the converter still operates at high resonance frequency $f_{sw,res}$ where the bottom plate losses become more dominant at low output power. Operation with dynamic

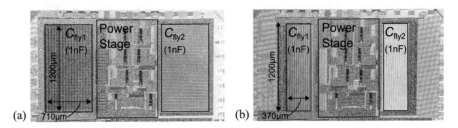

Fig. 5.5 Comparison of different implementation options for the flying capacitor C_{fly} (same 130 nm BCD technology): (**a**) with MIM capacitors; (**b**) with MOS capacitors

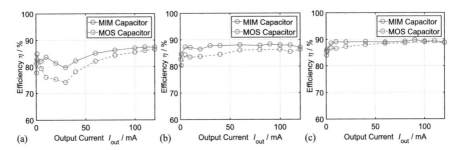

Fig. 5.6 Comparison of the measured converter efficiency for different implementation options of the flying capacitors at $V_{in} = 3.9$ V: (**a**) control with SwCR ($L = 10$ nH, $C_{out} = 10$ nF); (**b**) control with DOTM ($L = 10$ nH, $C_{out} = 100$ nF); (**c**) control with DOTM ($L = 39$ nH, $C_{out} = 100$ nF)

off-time modulation (DOTM) leads to a smaller efficiency drop of 3% (Fig. 5.6b), since the frequency modulation in DOTM scales down the bottom plate losses more effectively. Figure 5.6c shows that DOTM operation at lower switching frequencies $f_{sw,res}$ (higher inductance value L=39nH) leads to even smaller influence. With MOS capacitors an efficiency drop of only 2% has to be tolerated.

5.2 Implementation of Integrated Inductors

The main requirements for the integrated inductor in the resonant SC converter in this work are:

- Inductance value in the range of 10 nH and high inductance density for a small area solution.
- Usage of air-core inductors if possible in order to avoid magnetic core loss at the high switching frequencies of resonant SC converters.
- Low equivalent series resistance of $R_{ESR} < 400$ mΩ is required. Along with the requirements for the inductance value, this leads to a high-quality factor of $Q > 5.5$ necessary for resonant operation.

Integration of inductors is very challenging in standard CMOS since only planar designs are possible. However, because of their importance in power designs, there is a lot of research in the field of integrated inductors. In principle, there are two different approaches for realizing integrated inductors: with or without additional process steps of a CMOS chip. In the literature, many techniques have been demonstrated to improve the performance of integrated inductors. In [18–21], inductors with magnetic cores for higher inductance and quality factor are integrated on a separate interposer die, which is then wire-bonded or connected on the PCB. These solutions still suffer from relatively low inductance values. Moreover, parasitic impedances of the connections degrade the quality of the inductor. Core losses caused by the magnetic material are an additional drawback. In [22, 23], high-quality air-core inductors implemented as trace inductors within the package layer are introduced. On-die solenoid inductors with magnetic core directly on the same die as the active circuit are introduced in [24, 25]. High-quality factors can be achieved but the inductance values are limited to very small values. In [12, 13, 26, 27], small SMD air-core inductors are integrated in the package and placed on top of the die, which leads to a good overall performance. But still additional process steps for placing and bonding are necessary. Nevertheless, this solution was also investigated in this work [12, 13] (see Sect. 5.2.3).

All of the mentioned solutions are non-standard CMOS implementations, which are difficult and expensive to include in a standard CMOS process. Additionally, the used magnetic materials introduce excessive losses especially at high switching frequencies f_{sw}.

When using standard CMOS packaging techniques, the inductor can be directly formed with bond wires as a standard step in the CMOS fabrication process, shown in [28–31]. More details on this implementation technique are introduced in Sect. 5.2.1. Planar air-core inductors formed in the upper metal layers are traditionally implemented for RF applications [32–34] where small inductance values are integrated and higher-quality factors Q are reached in the GHz range. In recent years they have also been increasingly used in DCDC converters [35–40] where they are implemented mostly in inductive buck converters. There, the small inductance values lead to very high switching frequencies required for inductive PWM operation, which then lead to lower efficiencies. Resonant SC converters operate at much lower switching frequencies than inductive converters. Therefore, they can leverage the potential of planar air-core inductors which are discussed in Sect. 5.2.2.

5.2.1 Triangular Bond Wire Inductor

Bond wire inductors can be implemented in different geometric shapes. For the desired requirements (low ESR, high Q, and high inductance density), [29] shows that the triangular form is the best in terms of inductance density. Figure 5.7 shows the construction of an equilateral triangular bond wire inductance. The corners are

Fig. 5.7 Construction of
equilateral triangular bond
wire inductance [29]

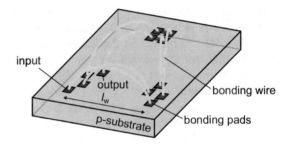

implemented with bonding pads, which have a higher resistivity than the bond
wires. This implementation offers low parasitic coupling due to the large distance
of the bond wires to the p-substrate. Additionally the area under the inductor can
be used for placing other converter components like capacitors or active circuits.
In order to evaluate the performance of this inductor implementation with the
specification needed for the resonant SC converter, the simplified heuristic equation
from Grover's inductance model [41] is used

$$L = \frac{\mu_0}{2\pi} \cdot 3l_\mathrm{w} \cdot \left(\ln\left(\frac{l_\mathrm{w}}{r_\mathrm{bond}}\right) - 1.5546 \right) \tag{5.2}$$

where l_w is the triangle side length and r_bond is the radius of the bond wire. This
simplified expression provides sufficient accuracy for a first design of the inductor
[29]. The equivalent series resistance (ESR) can be calculated from

$$R_\mathrm{ESR} = 3 \cdot \left(\frac{\rho l_\mathrm{w}}{A_\mathrm{eff}} + R_\mathrm{BP} \right) \tag{5.3}$$

where ρ is the resistivity ($\rho = 2.44 \cdot 10^{-8}\,\Omega\mathrm{m}$ (Au)) and R_BP is the resistance
of the bond pads required for the construction of the inductor. A_eff is the effective
cross-sectional area of the bond wire which results due to the skin effect. At higher
switching frequencies, the resistance of the conductor increases since the electrons
are forced toward the outer diameter of the conductor. Current flows mainly at
the "skin" of the conductor. The skin depth δ depends on the frequency f and is
defined as

$$\delta = \sqrt{\frac{\rho}{\pi \mu_0 f}}. \tag{5.4}$$

With Eq. 5.4, the effective area A_eff can be calculated

$$A_\mathrm{eff} = 2\pi r_\mathrm{bond}\delta - \pi \delta^2. \tag{5.5}$$

To evaluate the different inductor options, the quality factor Q is used. It is the ratio
of its inductive reactance to its resistance at a given frequency and is defined as

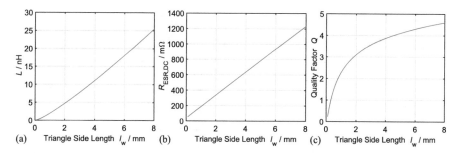

Fig. 5.8 Calculated parameters of a single triangular bond wire inductor (radius $r_{bond} = 12.5\,\mu m$) versus triangle side length l_w: (a) inductance L; (b) DC resistance $R_{ESR,DC}$; (c) quality factor Q of the inductor at a frequency of $f = 35\,MHz$

$$Q = \frac{Im(Z)}{Re(Z)}. \tag{5.6}$$

For a first approximation, the parasitic capacitance to the substrate is neglected here since there is a sufficiently large distance between the bond wires and the silicon substrate. Thus, the quality factor Q can be calculated with

$$Q = \frac{\omega L}{R_{ESR}}. \tag{5.7}$$

Figure 5.8 shows the results from Eqs. 5.2, 5.3 and 5.7 for a standard bond wire diameter of $d_{bond} = 25\,\mu m$. A design for an inductance value of $L = 10\,nH$ leads to a high ESR of $R_{ESR} = 600\,m\Omega$ which in turn leads to a low-quality factor Q of the inductor of $Q = 3.8$. A reduction of the ESR by increasing the inductance density with multiple turns is not really possible since the mutual inductance between the coils is relatively low because a wide spacing between the bond wires is required to prevent any short circuits. It also requires more bonding pads for construction which in turn introduce additional resistance. There may be also several challenges related to the mass production like reliability and reproducibility. In addition, wire bonding is more and more being replaced by flip-chip packaging technology where the construction of the inductor would cause additional costs. For all these reasons, wire bonding inductors are not further considered in this work.

5.2.2 Planar Inductor

On-chip planar inductors are usually made of the upper metal layers since they provide the thickest metal lines resulting in low ESR. This also leads to lower capacitive coupling due to the largest distance from the substrate, which improves the high-frequency performance.

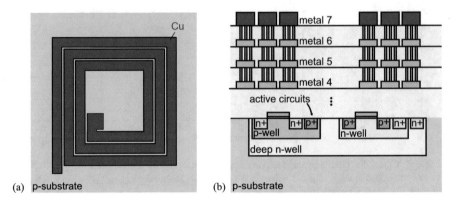

(a) p-substrate   (b) p-substrate

Fig. 5.9 Implementation of the planar inductor: (**a**) top view; (**b**) cross section

Fig. 5.10 Basic inductor
model

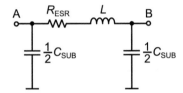

Basic Inductor Model

Figure 5.9 shows the construction of the integrated planar inductor. For a further
ESR reduction, the metal layers 4 to 6 are connected in parallel as depicted in
Fig. 5.9b. In the used process, the top metal layer (metal 7) is formed by 3-μm-
thick copper (Cu). The second top metal layer (metal 6) is formed by 900-nm-thick
aluminum (Al) layer and the remaining layers with a standard height of 380 nm. For
a low-ohmic connection, the inner winding end is directly connected with a bond
wire. This connection could be also realized in a flip-chip packaging technique.

There are many degrees of freedom for the design of a planar inductor. In order to
find a good design trade-off between the inductance value and parasitic components,
a model for the inductor has been implemented according to [42], (Fig. 5.10). The
inductance L consists of the self-inductance of the wire and the mutual inductances
between the windings of the inductor. The equivalent series resistance of the coil
is denoted by R_{ESR}. The substrate capacitance is represented by the capacitance
C_{SUB}. Half of C_{SUB} is connected to terminals A and B, respectively. The model
does not contain a coupling capacitance between the two terminals since the inter-
winding capacitances are small. Also, there is no return path under the inductor,
which would introduce significant coupling between the terminals. Substrate losses
due to the eddy currents, induced by the magnetic field of the coil, are not included
in this simple inductor model. They become dominant at much higher frequencies
than the used resonance frequency here. The measurement results show that they
can be neglected in the proposed operation range.

Fig. 5.11 Mutual inductances M between traces

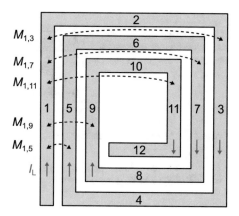

The inductance L can be calculated based on [43], expressed by

$$L = L_s + M_+ - M_- \qquad (5.8)$$

where L_s is the self inductance, M_+ the sum of the positive mutual inductances, and M_- the sum of the negative mutual inductances. M_+ results from the coupling between two conductors having currents in same directions (e.g., $M_{1,5}$, $M_{1,9}$), while M_- comes from the coupling between two conductors with currents in the opposite direction (e.g., $M_{1,11}$, $M_{1,7}$, $M_{1,3}$), shown in Fig. 5.11.

The self-inductance L_s can be calculated from Grover's equation for a straight rectangular wire [41, 43]

$$L_s = \frac{\mu_0 \cdot l_{winding}}{2\pi} \cdot \left(\ln\left(\frac{2l_{winding}}{w_{winding} + t_{winding}} \right) + 0.50049 + \frac{w_{winding} + t_{winding}}{3l_{winding}} \right) \qquad (5.9)$$

where $t_{winding}$ and $w_{winding}$ are the thickness and the width of the windings. $l_{winding}$ is the total length of all windings. The self-inductance L_s increases with an increasing total length $l_{winding}$ and with a decreasing width $w_{winding}$ or thickness $t_{winding}$.

The mutual inductances M_+ and M_- between two wires x and y can be calculated according to [43]

$$M_{x,y} = M(l_x + \lambda) - M(\lambda) \qquad (5.10)$$

where the different lengths l_x and λ are defined in Fig. 5.12. The mutual inductance $M(l)$ can be calculated with

$$M(l) = \frac{\mu_0}{2\pi} l_{winding} \left(\ln\left(\frac{l_{winding}}{GMD} \right) + \sqrt{1 + \left(\frac{l_{winding}}{GMD} \right)^2} - \sqrt{1 + \left(\frac{l_{winding}}{GMD} \right)^2} + \frac{GMD}{l_{winding}} \right) \qquad (5.11)$$

Fig. 5.12 Calculation of the mutual inductances M_+ and M_- between two segments x and y of the integrated planar inductor (top view)

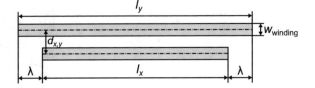

where GMD is the geometric mean distance which is defined with

$$GMD = \exp\left(\ln d_{x,y} - \frac{w_{\text{winding}}^2}{12 d_{x,y}^2} - \frac{w_{\text{winding}}^4}{60 d_{x,y}^4} - \frac{w_{\text{winding}}^6}{168 d_{x,y}^6} - \frac{w_{\text{winding}}^8}{360 d_{x,y}^8} - \frac{w_{\text{winding}}^{10}}{660 d_{x,y}^{10}} \right).$$
(5.12)

where $d_{x,y}$ is the distance of the track centers as shown in Fig. 5.12. The mutual inductance $M(l)$ is larger for a smaller distance $d_{x,y}$ between the segments since the magnetic coupling is enhanced.

For the equivalent series resistance R_{ESR} of the inductor, two effects have to be considered: the skin effect and the proximity effect. Yue and Wong [42] shows that the proximity effect can be neglected for windings that are on the same plane, which is the case for the proposed planar inductor. Hence, in the proposed model, only the resistance caused by the skin effect is included:

$$R_{\text{ESR}} = \frac{\rho \cdot l_{\text{winding}}}{w_{\text{winding}} \cdot t_{\text{winding,eff}}}$$
(5.13)

The resulting effective thickness of the conductor $t_{\text{winding,eff}}$ due to the skin effect can be calculated with

$$t_{\text{winding,eff}} = \delta \cdot (1 - e^{-t/\delta})$$
(5.14)

where δ is the skin depth, which was defined in Eq. 5.4. As the skin depth δ decreases with frequency, the series resistance R_{ESR} increases.

The parasitic capacitance C_{SUB} between metal and substrate is approximately proportional to the area occupied by the inductor and can be calculated with

$$C_{\text{SUB}} = l_{\text{winding}} \cdot w_{\text{winding}} \cdot \frac{\epsilon_0 \epsilon_{\text{r,oxide}}}{t_{\text{ox}}}$$
(5.15)

where l_{winding} is the total length of the inductor. t_{ox} denotes the thickness of the oxide between the inductor and the substrate. C_{SUB} limits the maximum achievable quality factor Q, which can be calculated from Eq. 5.7. For a small substrate capacitance C_{SUB}, the winding width w_{winding} and the number of winding N should be minimized. Smaller winding widths have also a positive impact on the mutual inductance M (Eq. 5.10) but, on the other hand, a negative impact on the series resistance R_{ESR}. A reduction of the number of winding N has a negative impact on the mutual inductance but lowers the series resistance R_{ESR}. In order to find an optimal design of the inductor, these relationships are compared against each other in the following subsection.

Design Trade-Offs

The main requirements for the integrated inductor in this work are obtained from Sect. 3.3. At the optimum design point, an inductance value of $L = 10\,\text{nH}$ together with a low equivalent series resistance of $R_{ESR} < 400\,\text{m}\Omega$ is required. This leads to a high-quality factor $Q > 5.5$ required for good resonance operation of the converter. There are several degrees of freedom for the design (winding width $w_{winding}$, number of winding N), which lead to trade-offs between inductance value L and resistive and capacitive parasitic components like R_{ESR} and C_{SUB}. An additional boundary condition is set by the geometry of the inductor, which was limited to $1500\,\mu\text{m} \times 1500\,\mu\text{m}$. For high magnetic coupling, the space $d_{x,y}$ between the track centers of the inductor has to be at the minimum. This is given by the rules for the metal layers in the used technology. This way, all geometric boundary conditions are given to find a final trade-off. Figure 5.13a shows the achievable inductance value L versus the possible number of windings N for different winding widths $w_{winding}$. Smaller winding widths lead to a higher number of windings and therefore to higher inductance values. However, it also increases the equivalent series resistance R_{ESR} (Fig. 5.13b), which lowers the achievable quality factor Q (Fig. 5.13c). With three inductor windings $N = 3$ and a winding width of $w_{winding} = 200\,\mu\text{m}$, the main requirements can be met. With these parameters, an inductance value of $L = 10\,\text{nH}$ with an DC series resistance of $R_{ESR} = 250\,\text{m}\Omega$ and a quality factor of $Q = 7.6$ at a frequency of $f = 35.5\,\text{MHz}$ is achieved.

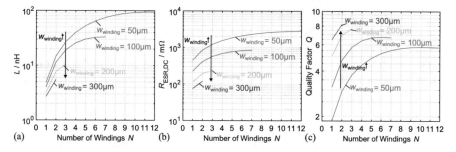

Fig. 5.13 Calculated planar inductor parameters for different number of windings N and winding widths $w_{winding}$: **(a)** inductance L; **(b)** DC resistance $R_{ESR,DC}$; **(c)** quality factor Q of the inductor at $f = 35.5\,\text{MHz}$

Fig. 5.14 Photograph of the
implemented inductor with
$N = 3$, $w_{\text{winding}} = 200\,\mu\text{m}$

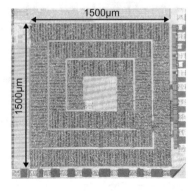

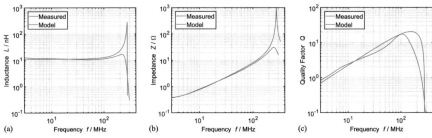

Fig. 5.15 Comparison between measurements and model for the implemented inductor ($N = 3$, $w_{\text{winding}} = 200\,\mu\text{m}$) versus frequency f: **(a)** inductance L; **(b)** impedance Z; **(c)** quality factor Q

Measurement Results

The proposed integrated inductor with $N = 3$ and $w_{\text{winding}} = 200\,\mu\text{m}$ was implemented in a 130 nm BCD process (Fig. 5.14). It is constructed with the metal layers 4 to 6 and the 3 μm-thick copper top metal layer as shown in Fig. 5.9b. Figure 5.15 shows the measurement results of the integrated inductor. Due to bond wires in the inductance test setup, a slightly higher inductance value of $L = 10.5$ nH is measured. The inductor has a DC resistance of $R_{\text{ESR,DC}} = 280\,\text{m}\Omega$ and a quality factor of $Q = 6$ at a frequency of $f = 35.5$ MHz. The results of the model are also shown in Fig. 5.15. In the relevant frequency range (<100 MHz), the model has a good agreement with the measurement results. At the peak resonance of the impedance Z and inductance L (Fig. 5.15b,c), there is a disparity between the calculations and measurement results. This can be expected since there is no modeling of the substrate losses. At the resonance peak, these substrate losses become more dominant since the occurring high currents lead to a large magnetic field, which leads to higher eddy currents in the substrate.

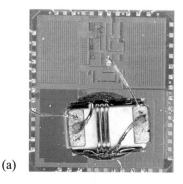 

(a)   (b)

Fig. 5.16 Different implemented options for off-chip inductor placed in the package: **(a)** 10 nH SMD inductor (Coilcraft 0603HP-10NXGLU); **(b)** usage of microfabricated on-silicon inductor [19, 44]

5.2.3 Off-Chip Inductors

Figure 5.16 depicts two different types of off-chip inductors, which are investigated in this work.

Figure 5.16a shows the implementation with a 10 nH SMD inductor from Coilcraft (0603HP-10NXGLU). The inductor is directly placed on top of the chip. In series production it could be co-integrated within the package. Short bond wires form the connection to the switching node and to the on-chip output capacitor. Compared to the integrated inductor, this solution has a much lower DC resistance of $R_{DC} = 48\,m\Omega$ and a quality factor of $Q = 20$ at a frequency of $f = 35.5\,MHz$.

Figure 5.16b shows a microfabricated on-silicon inductor, which is also placed on top of the chip. The component is actually a micro-transformer, which consists of a closed Co-Fe magnetic core and two coils, where one is on the primary side and the other is on the secondary side [19, 44]. In this work, only one coil is used, which provides an inductance value of $L = 11.9\,nH$. The fabrication technology is based on micro-electromechanical system (MEMS) fabrication technologies, which is easy to integrate with other systems and is well suited for further miniaturization and integration of DCDC converters. The structure is fabricated on top of an oxidized silicon wafer. After that, the transformer is removed from the silicon, with anodic dissolution which leads to a low profile height of $<150\,\mu m$. The inductor has a DC resistance of $R_{DC} = 150\,m\Omega$ and a quality factor of $Q = 10$ at a frequency of $f = 35.5\,MHz$.

5.2.4 Comparison of Different Inductor Options

In order to compare the different inductor options, the overall efficiency over the output current I_{out} of the converter is shown in Fig. 5.17. The inductor has

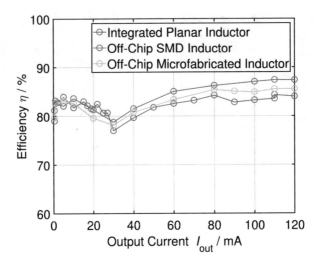

Fig. 5.17 Comparison of the measured converter efficiency for different implementation options of the inductor at $V_{in} = 3.9$ V

the most beneficial impact at higher output currents ($I_{out} > 50$ mA), where the converter operates with weakly damped resonant current waveforms. There, both off-chip inductors achieve higher efficiencies due to their better quality factors Q if compared with the integrated planar inductor. With the 10 nH SMD inductor from Coilcraft (0603HP-10NXGLU), an efficiency improvement of up to 4% is achieved. The microfabricated on-silicon inductor leads to an efficiency improvement of up to 2.3%. At lower output currents, the resonances are attenuated by SwCR, and the influence of the inductor on the efficiency decreases. When entering low-power mode ($I_{out} < 30$ mA), the converter acts as a conventional SC converter, and the control regulates the switching frequency instead of SwCR. In this case, the different inductor options have no longer any influence on the overall converter efficiency.

References

1. Andersen, T.M., et al.: A sub-ns response on-chip switched-capacitor DCDC voltage regulator delivering 3.7W/mm^2 at 90% efficiency using deep-trench capacitors in 32nm SOI CMOS. In: 2014 IEEE International Solid-State Circuits Conference Digest of Technical Papers (ISSCC), pp. 90–91 (2014). https://doi.org/10.1109/ISSCC.2014.6757351
2. Andersen, T.M., et al.: A feedforward controlled on-chip switched- capacitor voltage regulator delivering 10W in 32nm SOI CMOS. In: 2015 IEEE International Solid-State Circuits Conference – (ISSCC) Digest of Technical Papers, pp. 1–3 (2015). https://doi.org/10.1109/ISSCC.2015.7063076
3. Brunet, M., Kleimann, P.: High-density 3-D capacitors for power systems on-chip: evaluation of a technology based on silicon submicrometer pore arrays formed by electrochemical etching. IEEE Trans. Power Electron. **28**(9), 4440–4448 (2013). ISSN: 0885-8993. https://doi.org/10.1109/TPEL.2012.2233219

4. El-Damak, D., Bandyopadhyay, S., Chandrakasan, A.P.: A 93% efficiency reconfigurable switched-capacitor DCDC converter using on chip ferroelectric capacitors. In: 2013 IEEE International Solid-State Circuits Conference Digest of Technical Papers, pp. 374–375 (2013). https://doi.org/101109/ISSCC2013.6487776

5. McAdams, H.P., et al.: A 64-Mb embedded FRAM utilizing a 130nm 5LM Cu/FSG logic process. IEEE J. Solid State Circuits **39**(4), 667–677 (2004). ISSN: 0018-9200. https://doi.org/10.1109/JSSC.2004.825241

6. Jiang, J., et al.: A 2-/3-phase fully integrated switched-capacitor DCDC converter in bulk CMOS for energy-efficient digital circuits with 14% efficiency improvement. In: 2015 IEEE International Solid-State Circuits Conference – (ISSCC) Digest of Technical Papers, pp. 1–3 (2015). https://doi.org/10.1109/ISSCC.2015.7063078

7. Le, H., et al.: A sub-ns response fully integrated battery-connected switched-capacitor voltage regulator delivering 0.19W/mm^2 at 73% efficiency. In: 2013 IEEE International Solid-State Circuits Conference – (ISSCC) Digest of Technical Papers, pp. 372–373 (2013). https://doi.org/10.1109/ISSCC.2013.6487775

8. Butzen, N., Steyaert, M.: A 1.1W/mm^2-power-density 82%- efficiency fully integrated 3/1 Switched-Capacitor DCDC converter in baseline 28nm CMOS using stage outphasing and multiphase soft-charging. In: 2017 IEEE International Solid-State Circuits Conference (ISSCC), pp. 178–179 (2017). https://doi.org/10.1109/ISSCC.2017.7870319

9. Lutz, D., Renz, P., Wicht, B.: A 10mW Fully Integrated 2- to-13V-Input Buck-Boost SC Converter with 81.5% Peak Efficiency. In: 2016 IEEE International Solid-State Circuits Conference (ISSCC), pp. 224–225 (2016). https://doi.org/10.1109/ISSCC.2016.7417988

10. Lutz, D., Renz, P., Wicht, B.: A 120/230Vrms-to-3.3V micro power supply with a fully integrated 17V SC DCDC converter. In: ESSCIRC Conference 2016: 42nd European Solid-State Circuits Conference, pp. 449–452 (2016). https://doi.org/1.1109/ESSCIRC2016.7598338

11. Lutz, D., Renz, P., Wicht, B.: An integrated 3-mW 120/230-V AC mains micropower supply. IEEE J. Emerg. Sel. Top. Power Electron. **6**(2), 581–591 (2018). ISSN: 2168-6777. https://doi.org/10.1109/JESTPE.2018.2798504

12. Renz, P., et al.: A fully integrated 85%-peak-efficiency hybrid multi ratio resonant DCDC converter with 3.0-to-4.5V input and 50μA -to- 120mA load range. In: 2019 IEEE International Solid- State Circuits Conference – (ISSCC), pp. 156–158 (2019). https://doi.org/10.1109/ISSCC.2019.8662491

13. Renz, P., et al.: A 3-ratio 85% efficient resonant SC converter with on-chip coil for Li-Ion battery operation. IEEE Solid-State Circuits Lett. **2**(11), 236–239 (2019). ISSN: 2573-9603. https://doi.org/10.1109/LSSC.2019.2927131

14. Tong, T., et al.: A fully integrated battery-connected switched- capacitor 4:1 voltage regulator with 70% peak efficiency using bottom-plate charge recycling. In: Proceedings of the IEEE 2013 Custom Integrated Circuits Conference, pp. 1–4 (2013). https://doi.org/10.1109/CICC.2013.6658485

15. Castro Lisboa, P., et al.: General top/bottom-plate charge recycling technique for integrated switched capacitor DCDC converters. IEEE Trans. Circuits Syst. I Regul. Pap. **63**(4), 470–481 (2016). ISSN: 1558-0806. https://doi.org/10.1109/TCSI.2016.2528478

16. Butzen, N., Steyaert, M.S.J.: MIMO switched-capacitor DC–DC converters using only parasitic capacitances through scalable parasitic charge redistribution. IEEE J. Solid-State Circuits **52**(7), 1814–1824 (2017). ISSN: 1558-173X. https://doi.org/10.1109/JSSC.2017.2700009

17. Butzen, N., Steyaert, M.S.J.: Scalable parasitic charge redistribution: design of high- efficiency fully integrated switched-capacitor DC–DC converters. IEEE J. Solid-State Circuits **51**(12), 2843–2853 (2016). ISSN: 1558-173X. https://doi.org/10.1109/JSSC.2016.2608349

18. Sturcken, N., et al.: A 2.5D integrated voltage regulator using coupled-magnetic-core inductors on silicon interposer. IEEE J. Solid-State Circuits **48**(1), 244–254 (2013). ISSN: 0018-9200. https://doi.org/10.1109/JSSC2012.2221237

19. Dinulovic, D., et al.: Comparative study of microfabricated inductors/transformers for high-frequency power applications. IEEE Trans. Magn. **53**(11), 1–7 (2017). ISSN: 0018-9464. https://doi.org/10.1109/TMAG.2017.2734878

20. Dinulovic, D., et al.: On-chip high performance magnetics for point-of-load high-frequency DCDC converters. In: 2016 IEEE Applied Power Electronics Conference and Exposition (APEC), pp. 3097–3100 (2016). https://doi.org/10.1109/APEC.2016.7468306

21. Wu, H., et al.: Integrated transformers with magnetic thin films. IEEE Trans. Magn. **52**(7), 1–4 (2016). ISSN: 0018-9464. https://doi.org/10.1109/TMAG.2016.2515501

22. Burton, E.A., et al.: FIVR — fully integrated voltage regulators on 4th generation intel® core™ SoCs. In: 2014 IEEE Applied Power Electronics Conference and Exposition APEC 2014, pp. 432–439 (2014). https://doi.org/10.1109/APEC2014.6803344

23. Lambert, W.J., et al.: Package inductors for intel fully integrated voltage regulators. IEEE Trans. Compon. Packag. Manuf. Technol. **6**(1), 3–11 (2016). ISSN: 2156-3950. https://doi.org/10.1109/TCPMT2015.2505665

24. Krishnamurthy, H.K., et al.: A digitally controlled fully integrated voltage regulator with on-die solenoid inductor with planar magnetic core in 14-nm tri-gate CMOS. IEEE J. Solid-State Circuits **53**(1), 8–19 (2018). ISSN: 0018-9200. https://doi.org/10.1109/JSSC.2017.2759117

25. Wang, N., et al.: High-Q magnetic inductors for high efficiency onchip power conversion. In: 2016 IEEE International Electron Devices Meeting (IEDM), pp. 35.3.1–35.3.4 (2016). https://doi.org/10.1109/IEDM.2016.7838547

26. Kesarwani, K., Sangwan, R., Stauth, J.T.: A 2-phase resonant switched-capacitor converter delivering 4.3W at 0.6W/mm^2 with 85% efficiency. In: 2014 IEEE International Solid-State Circuits Conference Digest of Technical Papers (ISSCC), pp. 86–87 (2014). https://doi.org/10.1109/ISSCC.2014.6757349

27. Salem, L.G., Mercier P.P.: A single-inductor 7+7 ratio reconfigurable resonant switched-capacitor DCDC converter with 0.1-to-1.5V output voltage range. In: 2015 IEEE Custom Integrated Circuits Conference (CICC), pp. 1–4 (2015). https://doi.org/10.1109/CICC.2015.7338480

28. Abdulslam, A., et al.: A 93% peak efficiency fully-integrated multilevel multistate hybrid DC–DC converter. IEEE Trans. Circuits Syst. I Regul. Pap. **65**(8), 2617–2630 (2018). ISSN: 1549-8328. https://doi.org/10.1109/TCSI.2018.2793163

29. Piqué, G.V., Alarcón, E.: CMOS Integrated Switching Power Converters Springer, New York (2011). ISBN: 978-1-4899-8873-7. https://doi.org/10.1007/978144198843-0

30. Wens, M., Steyaert, M.: A fully-integrated 0.18μm CMOS DCDC step-down converter using a bondwire spiral inductor. In: 2008 IEEE Custom Integrated Circuits Conference, pp. 17–20 (2008). https://doi.org/10.1109/CICC.2008.4672009

31. Mostafa, M.A.L., Schlang, J., Lazar, S.: On-chip RF filters using bond wire inductors. In: Proceedings 14th Annual IEEE International ASIC/SOC Conference (IEEE Cat. No.01TH8558), pp. 98–102 (2001). https://doi.org/10.1109/ASIC.2001.954680

32. Long, J.R., Copeland, M.A.: The modeling, characterization, and design of monolithic inductors for silicon RF IC's. IEEE J. Solid-State Circuits **32**(3), 357–369 (1997). ISSN: 0018-9200. https://doi.org/101109/4557634

33. Koutsoyannopoulos, Y.K., Papananos, Y.: Systematic analysis and modeling of integrated inductors and transformers in RF IC design. IEEE Trans. Circuits Syst. II, Analog Digit. Signal Process. **47**(8), 699–713 (2000). ISSN: 1057-7130. https://doi.org/10.1109/82.861403

34. Dehan, M., et al.: Tapped integrated inductors: modelling and application in multi-band RF circuits. In: 2008 European Microwave Integrated Circuit Conference, pp. 234–237 (2008). https://doi.org/10.1109/EMICC.2008.4772272

35. Krishnamurthy, H.K., et al.: A 500MHz, 68% efficient, fully on-die digitally controlled buck voltage regulator on 22nm tri-gate CMOS. In: 2014 Symposium on VLSI Circuits Digest of Technical Papers, pp. 1–2 (2014). https://doi.org/10.1109/VLSIC.2014.6858438

36. Wens, M., Steyaert, M.S.J.: A fully integrated cmos 800-mw four-phase semiconstant ON/OFF-time step-down converter. IEEE Trans. Power Electron. **26**(2), 326–333 (2011). ISSN: 0885-8993. https://doi.org/10.1109/TPEL.2010.2057445

37. Kudva, S.S., Harjani, R.: Fully-integrated on-chip DCDC converter with a 450x output range. IEEE J. Solid-State Circuits **46**(8), 1940–1951 (2011). ISSN: 0018-9200. https://doi.org/10.1109/JSSC.2011.2157253

38. Kudva, S., Chaubey, S., Harjani, R.: High power-density hybrid inductive/capacitive converter with area reuse for multi-domain DVS. In: Proceedings of the IEEE 2014 Custom Integrated Circuits Conference, pp. 1–4 (2014). https://doi.org/10.1109/CICC.2014.6946053
39. Kim, W., Brooks, D., Wei, G.: A fully-integrated 3-level DCDC converter for nanosecond-scale DVFS. IEEE J. Solid-State Circuits **47**(1), 206–219 (2012). ISSN: 0018-9200. https://doi.org/10.1109/JSSC.2011.2169309
40. Amin, S.S., Mercier, P.P.: A fully integrated li-ion-compatible hybrid four-level DC–DC converter in 28-nm FDSOI. IEEE J. Solid-State Circuits **54**(3), 720–732 (2019). ISSN: 0018-9200. https://doi.org/10.1109/JSSC2018.2880183
41. Grover, F.W.: (1946). Inductance Calculations, Working Formulas and Tables. D. van Nostrand Company, New York
42. Yue, C.P., Wong, S.S.: Physical modeling of spiral inductors on silicon. IEEE Transactions on Electron Devices **47**(3), 560–568 (2000). ISSN: 0018-9383. https://doi.org/10.1109/16.824729
43. Greenhouse, H.: Design of planar rectangular microelectronic inductors. IEEE Trans. Parts Hybrids Packag. **10**(2), 101–109 (1974). ISSN: 0361-1000. https://doi.org/10.1109/TPHP.1974.1134841
44. Dinulovic, D., et al.: Microfabricated magnetics on silicon for point of load high-frequency DC–DC converter applications. IEEE Trans. Ind. Appl. **55**(5), 5068–5077 (2019). https://doi.org/101109/TIA20192921523

Chapter 6
Control of Resonant Switched-Capacitor Converters

In this chapter, the implementation details of the control for a resonant SC converter are shown. In Sect. 6.1, the control system implementation is presented, which comprises of two control options. Additional to that, experimental results are shown. A dynamic model of the resonant SC converter is introduced in Sect. 6.2. The modeling of the nonlinear switch conductance control loop is introduced in Sect. 6.3. It enables stability analysis and allows for an optimal design of the control loop, which is covered in Sect. 6.4.

6.1 Control System Implementation

Figure 6.1 shows the block diagram of the proposed multi-ratio ReSC converter with different control methods. It consists of the multi-ratio power stage, a switch controller, and an inner and outer control loop. The inner control loop implements different fine control techniques, depending on the implementation of the passives (see also Sect. 3.2).

Switch conductance regulation (SwCR) achieves a flat converter efficiency across the full load range due to effective scaling of the gate charge losses (see Sect. 3.2.2). It maintains small output voltage ripple, even with very small passives, and is hence suitable for fully integrated solutions without any external components. Multi-mode operation enables a transition into conventional SC mode for efficient light load operation with the switching frequency modulated by a hysteretic discrete-time control. This is described in Sect. 6.1.1. The second control option operates with dynamic off-time modulation (DOTM) and can be implemented with the same controller operating in hysteretic discrete-time control, which is discussed in Sect. 6.1.2. For highly efficient converter operation, all blocks of the control are designed to have minimum power consumption. The outer control loop is responsible for setting the appropriate conversion ratio depending on the input voltage as described in Sect. 6.1.3. A relaxation oscillator, which generates the clock signal φ required for all control methods, is introduced in Sect. 6.1.5.

6.1.1 Control Loop for Switch Conductance Regulation (SwCR) and SC Mode

The signals of Fig. 6.1 during operation of the SwCR and SC control are depicted in Fig. 6.2. First, the operation of the SwCR control loop is discussed. Figure 6.3 shows the implementation of the SwCR control loop. In SwCR mode, $\overline{SC\_mode} = 1$ applies. This enables the 8-bit counter, which is implemented as a synchronous binary up/down counter (on the right in Fig. 6.3). The output bits SwSel<0:7> change synchronously with the low-to-high transition of the input clock (driven from I17). A high level at the up/$\overline{down}$ input increments the counter value SwSel<0:7>, while a low level at up/$\overline{down}$ decrements the counter at every low-to-high transition of the clock input.

The converter uses hysteretic control and aims to maintain the feedback voltage V_{fb} within the hysteretic window ($V_{ref} - \Delta V$, $V_{ref} + \Delta V$), formed by two clocked comparators I1 and I2. The feedback voltage V_{fb} is generated from the output voltage via a resistive divider. The clocked comparators I1 and I2 maintain regulation by generating the up or down signals for incrementing or decrementing the switch selection bits SwSel<0:7>. Their clocked output signals are used to set and reset the RS flip-flops I4 and I5. I1 sends an up-pulse (up=1) when the feedback voltage V_{fb} falls below the lower boundary ($V_{ref} - \Delta V$) of the hysteresis (event (1) in Fig. 6.2). The up-signal from the comparator propagates through I11 before it is logically combined with the SC mode selection bit in I13 and with the clock signal φ in I16. The output of AND gate I16 is buffered by I17 to form the clock signal of the counter. The up-signal increases the SwSel<0:7> bits, which results in lower switch resistance R_{sw} of the power switches and, consequently, in lower equivalent output resistance R_{out}. The feedback voltage V_{fb} rises until it enters the hysteretic window (up=0) detected by comparator I1. A steady state is reached, and the counter stops counting up (between events (1) and (2) in Fig. 6.2). Similarly, comparator I2 sends a down-pulse when V_{fb} goes above the hysteretic window ($V_{ref} + \Delta V$) (event (2) in Fig. 6.2). The switch resistance R_{sw} of the power switches and therefore the equivalent output resistance R_{out} are increased, which leads to a decreasing output voltage. When changing the counting direction, the delay chain I17 ensures that the counting pulse at the clk input always occurs after a valid up/$\overline{down}$ input is present. Likewise, the clock signal φ propagates through a second delay chain I14 in order to synchronize with the up and down signals, which have to pass I4/I5, I9/I10, I11, and I13. Without the delay chain I14, a short glitch could occur during an up/down state change.

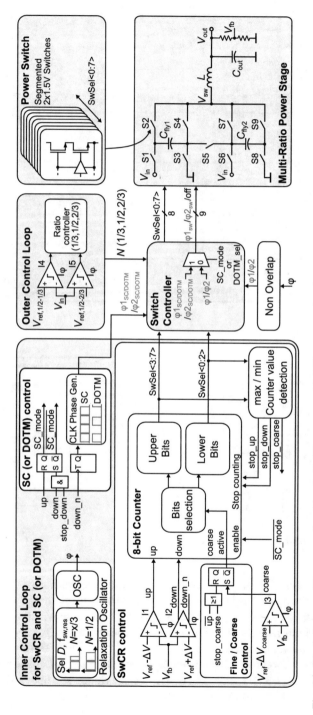

Fig. 6.1 Block diagram of the proposed multi-ratio ReSC converter with different control options

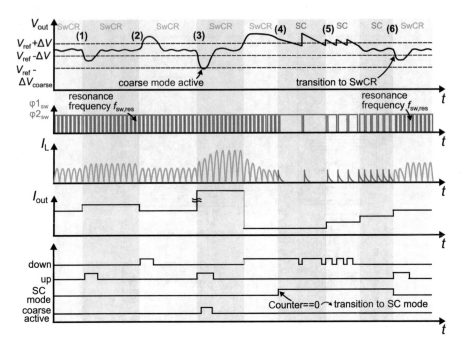

Fig. 6.2 Signals in SwCR and SC control

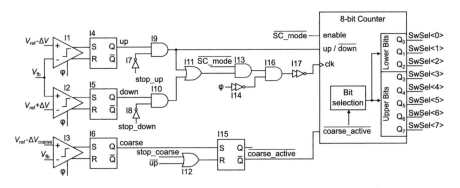

Fig. 6.3 Implementation of the SwCR control loop

In order to ensure fast transient response at large step-up load steps (event (3) in Fig. 6.2), a coarse mode has been implemented. An additional comparator I3 with a lower threshold $V_{ref} - \Delta V_{coarse}$ activates the coarse mode, in which only the five most significant bits SwSel<3:7> are controlled. The coarse mode is disabled when V_{fb} enters the hysteretic band or if all SwSel<3:7> bits are high. During large step-down load steps, the output voltage is limited since the converter uses a certain SC conversion ratio N (1/3, 1/2, 2/3). The output voltage cannot rise above $V_{in} \cdot N$. For this reason, no coarse control is used for this case.

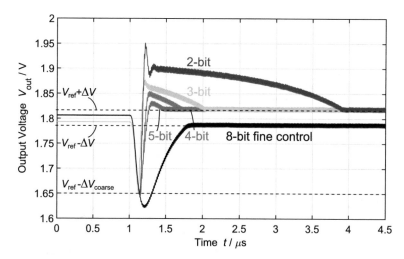

Fig. 6.4 Comparison of the control behavior for a coarse control with 2 bits, 3 bits, 4 bits, and 5 bits and a fine control with 8 bits, simulated transient responses to a 15 mA → 50 mA load step at $t = 1\,\mu s$

Figure 6.4 investigates the influence of the different number of bits in the coarse control. The response to a load step from 15 mA to 50 mA for a coarse control with 2 bits, 3 bits, 4 bits, and 5 bits is shown. Additionally, the transient response in fine control mode with 8 bits is plotted. In the first moment, a smaller number of bits leads to a faster reaction after the output voltage reaches the coarse threshold $V_{ref} - \Delta V_{coarse}$ at $t = 1.2\,\mu s$. But it also leads to larger overshoots of the output voltage, especially for 2-bit and 3-bit coarse control. A long response time of up to 2.9 μs (2 bits) is obtained, since the coarse control is disabled when the output voltage enters the hysteretic window ($V_{out} = V_{ref} - \Delta V$), while the counter value is still far from the required value. Thus, the settling time is even longer than the 8-bit fine control. The best trade-off between settling time and overshoot gives a coarse control that uses the five most significant bits.

The block entitled "max/min Counter value detection" in Fig. 6.1 consists of two subblocks, which observes all SwSel<0:7> bits. The minimum counter value detection, shown in Fig. 6.5a, generates the stop signals for both the coarse control (stop_coarse) and general up counting (stop_up). If all bits SwSel<0:7> are high, the maximum counter value detection block in Fig. 6.5b stops the counter from counting up in order to prevent an overflow. At this point the minimum switch resistance R_{sw} is reached, and the control cannot support higher output currents I_{out} anymore. The minimum counter detection block shown in Fig. 6.5b generates the stop_down signal, which prevents counter underflow by stopping the counter. This signal is also used to activate SC control in case of low output power.

Event (4) in Fig. 6.2 indicates the transition from SwCR control to SC control mode. A load step from high to low current leads to a rise in the feedback voltage V_{fb} and a decrementing of the counter value by the SwCR control. At low

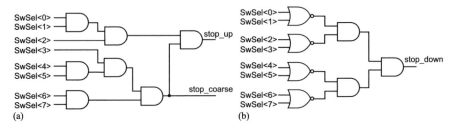

Fig. 6.5 Implementation of the maximum counter value detection (**a**) and the minimum counter value detection (**b**)

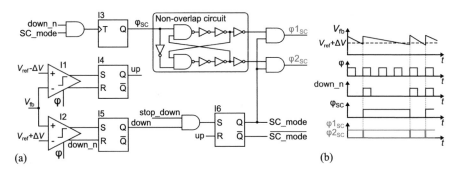

Fig. 6.6 (**a**) Implementation of the SC control loop; (**b**) signals in SC control with hysteretic discrete time control

output currents, the SwCR controller is not longer capable in controlling the output voltage, and the counter value reaches zero before the feedback voltage V_{fb} exceeds the hysteretic window. This is detected by the minimum counter detection block (stop_down=1), which activates the SC control mode (SC_mode=1) at the next down-pulse from comparator I2. As an advantage, the proposed method implicitly detects the load current instead of actually measuring it. This is the case in many approaches in order to switch to low-power operation [1]. No complex and power-consuming current sensor is required.

Figure 6.6a shows the implementation of the SC control loop. It reconfigures the clocked comparators I1 and I2 from the SwCR control loop with additional logic blocks and a non-overlap circuit. In SC control mode, the switching frequency is modulated by a hysteretic discrete-time control based on the output signal down_n of the clocked comparator I2. It compares the feedback voltage V_{fb} with the upper boundary $V_{ref} + \Delta V$ of the hysteretic window. The signals during operation are shown in Fig. 6.6b. When the feedback voltage V_{fb} is lower than the reference level $V_{ref} + \Delta V$ at the rising edge of the clock signal φ, down_n at the output of comparator I2 is set to high. This changes the output φ_{SC} of the T-flip-flop, which is converted into two non-overlap clock signals $\varphi1_{SC}$ and $\varphi2_{SC}$. In this way, the frequency modulation adjusts the equivalent output resistance of the converter

(Eq. 2.25) in order to maintain the output voltage close to the upper threshold of the hysteresis window at event (5) in Fig. 6.2.

When the load current increases and the SC mode is no longer capable of supplying the output current, the voltage V_{fb} drops below the lower threshold $V_{ref} - \Delta V$ of the hysteresis window. This is detected by comparator I1, and its up-signal is used to return to SwCR mode (event (6) in Fig. 6.2).

6.1.2   Control Loop for Dynamic Off-Time Modulation (DOTM)

For the dynamic off-time modulation (DOTM), resonant pulses have to be generated where the required pulse widths depend on the resonance frequency $f_{sw,res}$ defined by the effective flying capacitance C_{fly} and the inductance L as derived in Eqs. 3.1 and 3.3. The effective value of the flying capacitance, either as series or as parallel connection of C_{fly1} and C_{fly2}, depends on the conversion ratio of the ReSC voltage converter. The control loop modulates the time distance between the resonant pulses.

This loop can be implemented as a hysteretic discrete-time control introduced for the SC control in Sect. 6.1.1. Figure 6.7a shows the implementation of the DOTM control loop. The clocked comparator I1 together with the T-flip-flop I2 and the DOTM pulse generation block forms the hysteretic control loop. The output φ_{DOTM} of the T-flip-flop is the control signal in DOTM mode. It defines the duration between the resonant pulses, which are generated by the following pulse generation circuit. The pulses are triggered by the edges of φ_{DOTM}, as shown in Fig. 6.7b. The pulse duration is tunable, not only to compensate process variations but also to support different external inductances and capacitances for test purposes. The delay tuning is based on current-starved inverters I3 and I8 and on capacitive loads C_{1a}, C_{1b}, C_{2a}, and C_{2b} at the output of the inverters. For the pulse generation, only the falling output edge of I3 and I8 has to be current starved. The rising edge, which recharges the load capacitors C_{1a}, C_{1b}, C_{2a}, and C_{2b} to V_{DD}, sees a much faster hard-switching transition. This is required to enable a nearly continuous series of pulses at $\varphi1_{DOTM}$ and $\varphi2_{DOTM}$ as shown in Fig. 6.7b. To support experiments with external inductors and capacitors, pulse widths of up to 70 ns have to be generated. Since this leads to a very slow falling edge at the output of the current-starved inverter, Schmitt trigger inverters I4 and I9 are used. This generates a sharp edge at the output for following logic gates.

First, the function for the 1/2 ratio (assuming ratio_1/3 = ratio_2/3 = 0) is explained, followed by the other ratios $(x/3)$. At the rising edge of φ_{DOTM}, both inputs of the AND gate I6 are high, and $\varphi1_{DOTM}$ gets high. Both capacitors C_{1a} and C_{1b} are in parallel and get slowly discharged by the current-starved inverter I3 with the current I_{starve}. When the voltage V_{C1} reaches the lower switching limit of the Schmitt trigger inverter I4, both I4 and I5 toggle. The connected input of the AND gate I6 changes to low and causes a falling transition of $\varphi1_{DOTM}$ again. This way, a

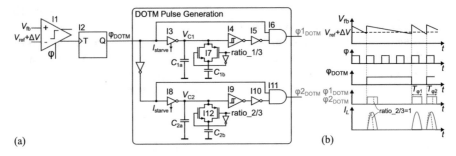

(a)   (b)

Fig. 6.7 (a) Implementation of the DOTM control loop; (b) signals in DOTM control with hysteretic discrete-time control

pulse on $\varphi1_{DOTM}$ is generated, triggered by the rising edge of the input φ_{DOTM}. At the falling edge of φ_{DOTM}, the procedure is vice versa, the capacitors C_{1a} and C_{1b} are recharged to V_{DD}, and a pulse on $\varphi2_{DOTM}$ is created by the lower path consisting of I8 to I11.

The DOTM clock signals $\varphi1_{DOTM}$ and $\varphi2_{DOTM}$ are supposed to have the same pulse width when the converter operates in the 1/2 ratio. For $C_{fly1} = C_{fly2} = 1\,nF$ and $L = 10\,nH$, this results in a pulse width of 14 ns for both phases. In this case, both control signals ratio_1/3 and ratio_2/3 are low, and the capacitances C_{1b} and C_{2b} are connected in parallel to C_{11} and C_{2a} accordingly. During operation, the capacitances are discharged from V_{DD} to 0.3 V, which is the lower switching limit of the Schmitt trigger inverters I4 and I9. Due to this large voltage swing, transfer gates I7 and I12 are used to keep the capacitances well-connected to the outputs of the current-starved inverters. The pulse width of the clock phase $\varphi1_{DOTM}$ can be calculated from

$$T_{\varphi1} = T_{d,I4} + T_{d,I5} + \frac{C_{1a,b} \cdot (V_{DD} - V_{th,low,I3})}{I_{starve}} \quad (6.1)$$

where $T_{d,I4}$ and $T_{d,I5}$ are the propagation delays of inverters I4 and I5 and $C_{1a,b}$ is the effective capacitance value at node V_{C1}. The pulse width of $\varphi2_{DOTM}$ can be calculated in the same way.

When the ReSC converter operates in ratio 2/3, the control signal ratio_2/3 is high. Thereby C_{2b} is disconnected from the output of the current-starved inverter I8. This leads to a smaller effective capacitance value at node V_{C2} and therefore to a shorter propagation delay and shorter pulse width of $\varphi2_{DOTM}$ as required for the correct operation of the ReSC converter. The capacitances are designed to achieve the desired pulse widths of 14 ns when both capacitances form the load of the current-starved inverter and 10 ns when the additional capacitance C_{2b} is disconnected. This applies to both current-starved inverters I3 and I8. Due to the linear relationship between pulse width $T_{\varphi2}$ and load capacitance $C_{2a,b}$, the current I_{starve} has to be trimmed only once to tune the pulse width to the resonance frequency of each phase of the ReSC converter. The adaption of the pulse width to

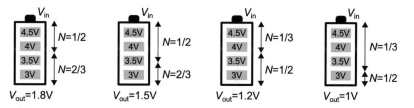

Fig. 6.8 Selection of the conversion ratio N depending on the input voltage V_{in} and output voltage V_{out}

the different conversion ratios is done by adding the capacitances, while the bias current I_{starve} can remain unchanged. In the same way, the pulse width $\varphi 1_{DOTM}$ is adapted to be shorter in ratio 1/3 with the control signal ratio_1/3 = 1.

6.1.3 Outer Control Loop

The different conversion ratios N of the ReSC converter offer a lossless coarse control of the output voltage. The outer control loop is responsible for the choice of the appropriate conversion ratio N depending on the input voltage V_{in} and the output voltage V_{out}. It is active in all fine control modes, SwCR, SC mode, and DOTM. Figure 6.8 illustrates which ratios are necessary for the conversion of input voltages within the full Li-Ion range to different output voltages ranging from 1.8 V to 1 V. For output voltages of 1.8 V–1.5 V, only ratios 1/2 and 2/3 are required, while lower output voltages of 1.2 V and 1 V utilize the ratios 1/3 and 1/2.

Figure 6.9 shows the implementation of the outer control loop, based on resistive dividers. Two clocked comparators I1 and I2 together with the RS flip-flops I3 and I4 are used for detection of the input voltage V_{in}. The ratio controller evaluates the comparator output signals A and B and sets the appropriate ratio (ratio_1/3, ratio_1/2, ratio_2/3). The switch controller, shown in Fig. 6.1, controls the power switches in the power stage accordingly.

For efficient low-power operation, each of the two resistive dividers R_1-R_5 is designed with a total resistance of 500 kΩ. Since the rate of change of the input voltage V_{in} in battery applications is slow, the buffer capacitors C_1–C_3 are set to 200 fF. This improves the noise immunity and prevents multiple triggering of the comparators when the input voltage gets close to the reference voltages. As described in Sect. 3.1, changing of the conversion ratio results in different resonance frequencies and pulse widths. Therefore, the ratio information is also used by the oscillator, which generates the clock signal for the ReSC power stage (see Sect. 6.1.5). The response to an input voltage step is shown in Sect. 6.1.6, which proves the functionality of the outer control loop.

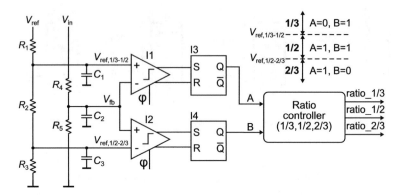

Fig. 6.9 Implementation of the outer control loop for changing the conversion ratio N

6.1.4 Clocked Comparators

Several comparators are required for the implementation of the control loops for sensing the output voltage V_{out} and the input voltage V_{in} of the converter (Sects. 6.1.1 and 6.1.3). Clocked regenerative comparators (also called dynamic comparators or latch-type comparators) are well suitable in this design because of their low power consumption. They enable high-speed operation due to positive feedback. Clocked comparators play a crucial role in the design of Flash-ADCs [2, 3] and are also often applied to read the content of different memory types [4].

Figure 6.10a shows the well-known conventional latch-type comparator, which was introduced in [5] and investigated for minimum offset in [6]. It consists of a clocked differential pair (MN2-MN3) and two cross-coupled inverters (MN4-MP2 and MN5-MP3), which provide the positive feedback. Additional to that, three reset switches (MN1, MP1, and MP4) are required. During the reset phase when $\varphi = 0$, both output nodes V_{out+} and V_{out-} are pulled to V_{DD} for a defined start condition. With $\varphi = V_{DD}$, the sensing or regeneration phase starts, and the two output nodes V_{out+} and V_{out-} start to discharge with different rates related to the corresponding input voltages (V_{in+} and V_{in-}). Depending on the polarity of the voltage between the output nodes, the inverter latch will flip in one or the other direction. There is only a current flow in the sensing phase but no static current consumption. Fast transients lead to kickback currents drawn from the input. Different impedances at the inputs can lead to a disturbance of the input voltage, which degrades the accuracy. The common mode kickback current from turning on MN1 is the dominating part since it is drawing its drain current from the gate-source capacitances of MN2 and MN3.

The kickback current can be reduced with the comparator shown in Fig. 6.10b [7] where the input devices MN1 and MN2 are clocked through their drain path rather than their source path. Therefore, transistors MN5 and MN6 are introduced to reset and enable the latch. However, this results in a higher input offset since the

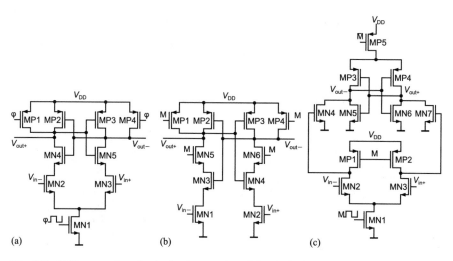

Fig. 6.10 Different options for the implementation of the clocked comparators: (**a**) conventional latch-type comparator [5]; (**b**) latch-type comparator for lower kickback noise [7]; (**c**) double-tail latch-type comparator [8]

transistors MN1 and MN2 now operate in the triode region (depending on the input voltage) during the sensing phase [7].

Figure 6.10c shows a double-tail latch-type comparator [8] as an alternative implementation. It uses one tail for the input stage (MN2-MN3) and another one for the latch stage (MN5-MP3 and MN6-MP4). Due to fewer stacked transistors, this topology can operate at lower supply voltages V_{DD}. The intermediate stage formed by MN4 and MN7 provides also an additional shielding between input and output, which leads to lower kickback noise.

For comparison, all comparator options in Fig. 6.10 are implemented in a 130 nm BCD technology as part of this work. For comparison, the differential input pairs of the different options are designed with the same size. Table 6.1 shows the measurement results. The conventional latch-type comparator achieves the lowest energy per conversion, offset voltage, and the shortest delay.

The energy consumption is the most critical parameter for selection of the comparator in this converter design and is mainly determined by the number of used transistors. Low energy consumption is required for efficient low-power operation of the converter. Therefore, the conventional latch-type comparator is selected. In the proposed application, a slightly higher kickback noise is not critical since the impedances at the inputs of the comparators are similar, which compensates the influence of the common mode kickback current. Additional to that, small capacitors are placed at the input for a further reduction of the kickback noise. In the final implementation, shown in Figs. 6.3 and 6.6, a RS flip-flop follows the clocked comparator. This is required since its outputs V_{out+} and V_{out-} are pulled to V_{DD} for half of the clock cycle.

Table 6.1 Measurement results of different clocked comparator options

	Conventional latch-type [5], Fig. 6.10a	Latch-type (low kickback) [7], Fig. 6.10b	Double-tail [8], Fig. 6.10c
Energy per conversion/pJ	0.2	0.29	0.38
Input-referred offset voltage/mV	5	8.5	8
Delay/ps (@ $\Delta V_{in} = 20$ mV)	570	915	940
Transistor count	9	10	12

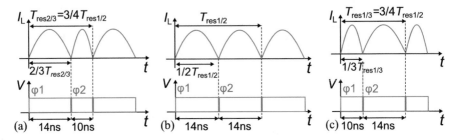

Fig. 6.11 Principle waveforms of $\varphi 1$ and $\varphi 2$ for the nominal values of the flying capacitors $C_{fly1} = C_{fly2} = 1$ nF and the inductance $L = 10$ nH: (**a**) 1/2 ratio; (**b**) 2/3 ratio; (**c**) 1/3 ratio

6.1.5 Oscillator

The ReSC converter requires a constant clock signal φ at the resonance frequency of the respective conversion ratio, as shown in Sect. 3.1. The clock signal φ is also fed to all clocked comparators used for the different control options (SwCR, SC mode, and DOTM); see Fig. 6.1. In SwCR mode, the two clock signals $\varphi 1$ and $\varphi 2$ are generated from φ with a non-overlap circuit, which are directly used to drive the power switches. In the 1/2 ratio, the two flying capacitors are always in parallel. For the 1/3 and 2/3 ratio, the capacitors are connected in series during one of the phases, resulting in a factor 4/3 higher resonance frequency and different duty cycle. This leads to resonance frequencies $f_{sw,res}$ of 35.5 MHz (1/2 ratio) or 47.5 MHz (x/3 ratio) with the nominal values of the flying capacitors $C_{fly1} = C_{fly2} = 1$ nF and the inductance $L = 10$ nH.

The waveforms are illustrated in Fig. 6.11.

The oscillator has to be capable to switch between the required resonance frequencies and different duty cycles depending on the conversion ratio. This is achieved with a current-controlled relaxation oscillator according to Fig. 6.12. It consists of a current source I_{osc}, which is used to charge two capacitors C_1 and C_2 alternately. The two capacitors can be matched very well in the manufacturing process. When the output φ is high, the transistors MN2 and MN3 turn on. C_1 is shorted to GND, while the current I_{osc} charges C_2. When V_{C2} reaches the switching threshold V_{ref} of comparator I2, its output V_{o2} is pulled down, which toggles the NAND latch. The output $\bar{\varphi}$ turns high and, consequently, the output φ low. MN4

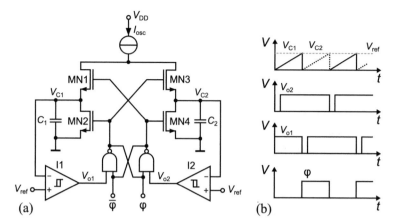

Fig. 6.12 (a) Operation principle of a two phase relaxation oscillator; (b) simplified waveforms

shorts C_2, which is thereby discharged. The comparator I2 toggles back to high and the NAND latch keeps its state. Now capacitance C_1 is charged with the current I_{osc} via transistor MN1 until its voltage V_{C1} reaches the switching threshold of comparator I1. The comparator is based on the comparator with internal hysteresis discussed in [9] implemented with a PMOS differential input stage supporting low threshold voltages. The employed comparator also comprises an internal hysteresis to improve the noise performance and prevent multiple triggering, when the input voltage gets close to the reference voltage V_{ref} applied at the other input of the comparator. It offers short propagation delay of $T_{d,comp} = 3.5$ ns. The hysteresis is set to $V_{hys} \pm 110$ mV around the reference voltage V_{ref}. The two complementary outputs of the NAND latch provide a two-phase clock signal. To adapt the duty cycle of the clock for the different conversion ratios of the ReSC converter, there are two possibilities: one is to apply different reference voltages V_{ref} to the comparators, and the other is to increase the current I_{osc} for one of the capacitances. For this implementation, the second option is chosen as it can readily be implemented.

Figure 6.13 shows the schematic of the proposed relaxation oscillator, which supports variable duty cycles. The charging current I_{osc} is generated from a bias current I_{bias} with a PMOS cascode current mirror formed by MP1–MP4. For testing purposes the bias current is supplied externally. For the oscillator design, 1.5 V transistors are used. A cascode current mirror is preferred over a simple current mirror since it is less susceptible to voltage changes at the drain of MP3. The matching of the currents declines for drain voltages greater than 500 mV at MP3. At higher voltages, the drain-source voltages of both transistors, MP3 and MP4, decrease too much to operate within the saturation region. For the design of the relaxation oscillator, this sets the upper limit of the switching threshold of the comparators. For reliable operation, a current of $I_{osc} = 20\,\mu A$ is used. With a switching threshold of $V_{ref} = 400$ mV, the required values for the capacitors C_1 and C_2 can be calculated with

$$C_1 = C_2 = \frac{I_{\text{osc}} \cdot \left(\dfrac{1}{2 f_{\text{sw,res}}} - T_{\text{d,comp}} \right)}{V_{\text{ref}} + V_{\text{hys}}} \tag{6.2}$$

where $T_{\text{d,comp}}$ is the comparator delay and V_{hys} is the hysteresis of the comparator. To obtain a resonance frequency of $f_{\text{sw,res}} = 35.5\,\text{MHz}$, the capacitors C_1 and C_1 are implemented with $C_1 = C_2 = 415\,\text{fF}$. MIM capacitors are used, since they offer the most constant capacitance value independent of the stored voltage and charge. A shorter pulse width and higher switching frequency, required for the ratios 2/3 and 1/3, are created with an additional current $I_{\text{osc,add}}$, which can be added to the oscillator current I_{osc} when charging the capacitor C_2. This current is generated with the additional current mirror MP7 and MP8. The mirror ratio of MP8 to MP2 is designed to provide the required pulse widths without adapting the bias current I_{bias}. The transistor MN6 turns the additional current on when the converter is operated

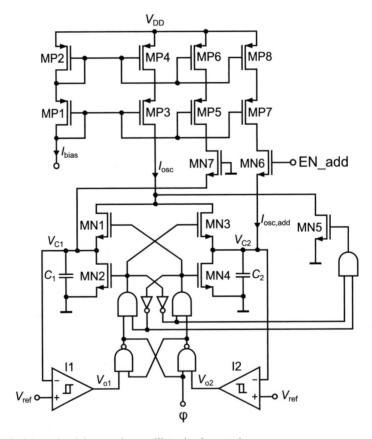

Fig. 6.13 Schematic of the complete oscillator implementation

in the 2/3 or 1/3 ratio. An additional dummy branch of MP5, MP6, and MN7 is inserted to achieve best matching of the parasitic capacitances at the nodes V_{C1} and V_{C2}.

During the transition of the basic NAND latch, all four NMOS transistors MN1–MN4 are conductive for a short time (see Fig. 6.12a). Assuming that C_1 has just been charged and C_2 is discharged, some charge is transferred from C_1 to C_2 during the transition time of the NAND latch. Thereby the charging of C_2 does not start at 0 V as desired, but at around 100 mV. To avoid this, the NAND latch is extended by a non-overlap circuit formed by two AND gates and two inverters. This ensures that MN3 and MN1 are never turned on at the same time and no charge transfer can occur between the two capacitances. While this improves the performance of the oscillator, it compromises the operation of the current mirror. As long as MN1 and MN3 are turned off, there is no path for the current I_{osc}. Especially at high oscillator currents, the parasitic capacitance at the common drain of MN1 and MN3 is quickly charged with the oscillator current, which cuts the current mirror off. If now MN1 or MN3 is turned on, the current mirror needs some time to start over again and provide a settled I_{osc}. Moreover, the charge stored on the parasitic capacitance is transferred to the corresponding capacitance C_1 or C_2. The charging of the capacitance with I_{osc} starts at around 40 mV at nodes V_{C1} and V_{C2}. To avoid these negative effects, an alternative current path is provided with MN5, which is turned on with another AND gate during the time slot when MN1 and MN3 are both turned off. This finally leads to the desired initial condition for the charging of the capacitance C_1 or C_2 from nearly 0 mV. The charge injection at the turn-off of MN5 is negligible.

For the same reason, the additional current $I_{osc,add}$ is directly guided to the top plate of C_2 and is always flowing when the ReSC converter is operated in the 2/3 or 1/3 ratio. The resistance of MN4 is small enough to guide the current $I_{osc,add}$ to GND with a very low voltage drop of less than 2 mV during the discharging phase of C_2.

6.1.6 Experimental Results

SwCR and SC Mode Control

Figures 6.14 and 6.15 show transient load step responses for the fully integrated converter option with SwCR and SC mode control. A load step from 10 mA to 50 mA (30 ns rise time), when the control changes the operation mode from SC mode to resonant SwCR, is shown in Fig. 6.14a. The proposed control enables a very fast transition from SC mode to SwCR, which is activated when the output voltage falls below the lower threshold $V_{out} < V_{ref} - \Delta V$. This leads to a small voltage drop of 80 mV ($\Delta V_{out} = 4.4\%$) and fast settling within 170 ns. Figure 6.14b shows the transient response for a larger load step from 30 mA to 120 mA (50 ns rise time) within the resonant SwCR operation mode. The output voltage drop triggers the coarse mode ($V_{out} < V_{ref} - \Delta V_{coarse}$) leading to a fast response time of 250 ns and a small voltage drop of 100 mV ($\Delta V_{out} = 5.5\%$). The coarse mode is stopped when V_{out} enters the hysteretic band as described in Sect. 6.1.1. Disabling of coarse mode leads to a higher voltage drop of $\Delta V_{out} = 10\%$ and significant longer response time of 2.5 µs.

The transient response for a load step from 50 mA to 10 mA is shown in Fig. 6.15a. The converter is in SwCR control, while the load step leads to a rise in the output voltage and a decrementing switch conductance counter value (see Fig. 6.3). At $t = 1.5$ µs, the counter reaches its minimum value and activates the SC control mode. The SC mode modulates the switching frequency in order to bring the output voltage back into the hysteretic window. A response time of 550 ns and an overshoot of 100 mV ($\Delta V_{out} = 5.5\%$) are achieved. Figure 6.15b shows the response to a load step within the SwCR control from 90 mA to 40 mA. An overshoot of 60 mV ($\Delta V_{out} = 3.3\%$) together with a response time of 1.2 µs is achieved. Since no coarse control is used for step-down load steps (see Sect. 6.1.1), slightly longer response times are obtained compared with the step-up load step responses in Fig. 6.14.

The transient response to a reference voltage V_{ref} step is shown in Fig. 6.16a. The converter operates at a input voltage of 3.9 V, an output voltage of 1.5 V, and an output current of 60 mA. After a reference voltage step, the converter settles quickly at the new output voltage of 1.8 V within 500 ns.

A transition between two conversion ratios is shown in Fig. 6.16b where the input voltage changes from 3.6 V to 4 V. The long rise time of 80 µs, caused by the measurement setup, is tolerable as the Li-Ion battery voltage also changes relatively slowly in the application. The converter operates at $V_{out} = 1.8$ V and $I_{out} = 80$ mA. For input voltages $V_{in} < 3.85$ V, the converter operates in ratio 2/3. The quantization error of the SwCR control leads to the gradations in the output voltage between 0 µs $< t < 50$ µs. These become larger with increasing switch resistances in the input voltage range 3.6 V $< V_{in} < 3.85$ V (see Fig. 3.10 in Sect. 3.2.2). During the

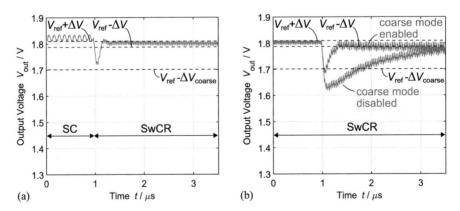

Fig. 6.14 Measured transient load step response for the fully integrated converter option with SwCR and SC mode control, $V_{in} = 4$ V, $V_{out} = 1.8$ V: **(a)** transient response to a 10 mA → 50 mA load step at $t = 1$ µs; **(b)** transient response to a 30 mA → 120 mA load step at $t = 1$ µs

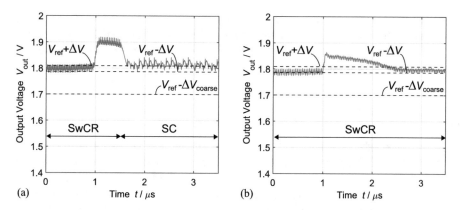

Fig. 6.15 Measured transient load step response for the fully integrated converter option with SwCR and SC mode control, $V_{in} = 4$ V, $V_{out} = 1.8$ V: **(a)** transient response to a 50 mA $\rightarrow$ 10 mA load step at $t = 1$ μs; **(b)** transient response to a 90 mA $\rightarrow$ 40 mA load step at $t = 1$ μs

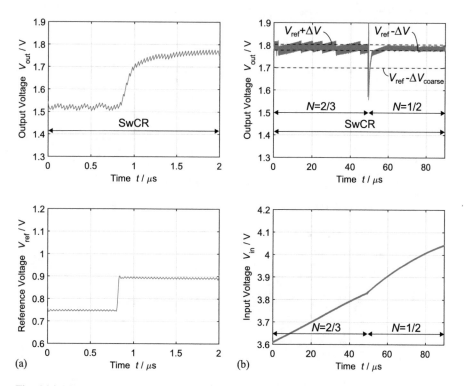

Fig. 6.16 Measured transient load response for the fully integrated converter option with SwCR and SC mode control: **(a)** measured transient response to a reference voltage V_{ref} step for $V_{in} = 4$ V, $I_{out} = 60$ mA resulting in $V_{out} = 1.5$ V $\rightarrow$ 1.8 V; **(b)** transient response to an input voltage transition, $V_{out} = 1.8$ V, $I_{out} = 80$ mA

transition of the conversion ratio from ratio 2/3 to 1/2 at $V_{in} = 3.85$ V, the output voltage settles within 2 μs with a maximum undershoot of 190 mV. The undershoot results from the high resistance value required for regulation at the end of ratio 2/3. This must first be reduced by the SwCR in ratio 1/2.

DOTM Mode Control

In DOTM control, a small 10 nH SMD inductor and a 100 nF output capacitor are used due to the high and non-continuous current pulses, which would lead to a large output voltage ripple (see Sect. 3.2.1). Figure 6.17a shows the transient load step response. Due to the fast control loop and the relatively large external output capacitor, a 30 mA to 120 mA load step at $t = 1$ μs (50 ns rise time) has a negligible voltage drop of $\Delta V_{out} < 1\%$. Figure 6.17a also shows the clock signals $\varphi 1_{DOTM}$ and $\varphi 2_{DOTM}$. When the load step occurs, the time distance between the resonance bursts is reduced accordingly. Figure 6.17b shows the transition between the conversion ratios 2/3 and 1/2 related to an input voltage transition from 3.5 V to 4.1 V. There,

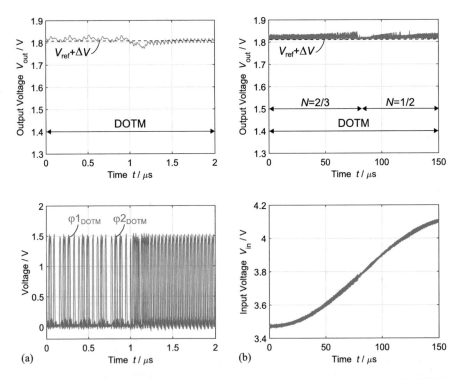

Fig. 6.17 Measured transient load response for the highly integrated converter option with DOTM control, $V_{in} = 4$ V, $V_{out} = 1.8$ V, $C_{out} = 100$ nF (external), $L = 10$ nH (external): **(a)** transient response to a 30 mA → 120 mA load step at $t = 1$ μs; **(b)** transient response to an input voltage transition, $I_{out} = 50$ mA

the converter operates at an output voltage of 1.8 V and a output current of 50 mA. For input voltages $V_{in} < 3.85$ V, the converter operates in ratio 2/3. Lower switching frequency, required for regulation at higher input voltages (3.5 V $< V_{in} < 3.85$ V), leads to a slightly higher output voltage ripple (see Sect. 3.2.1). Due to the fast control loop and external 100 nF output capacitor, the transition between the ratios happens smoothly without any undershoot or overshoot of the output voltage.

6.2 Plant Transfer Function of Resonant SC Converters

In Sect. 2.3.2, the equivalent output resistance model is introduced (Fig. 2.16), which models the capacitive charge transfer losses, conduction losses, and static load behavior of the converter. In this section, the static model is expanded to include the dynamic response to both input voltage and output current changes. The model is later on used to design the control loop according to stability requirements (Sect. 6.3.1). There are various approaches for modeling of the dynamic behavior of SC converters in the literature [10–16]. However, they are often based on very complex calculations and do not apply the easy-to-use charge multiplier framework from [16]. Moreover, they are not directly compatible with the introduced equivalent output resistance model in Sect. 2.3.2. Dynamic modeling of resonant SC converter has not been covered in the literature so far. With an inductor as an additional energy storage element, the calculations in [10–16] would become even more complex. In Sect. 6.2.1, a dynamic model for the SC operation mode is introduced, which relies on the equivalent output resistance. Based on these considerations, a dynamic model for the resonant SC converter is introduced in Sect. 6.2.2.

6.2.1 SC Mode

The analysis of the behavior of SC filters dates back to the 1970s [17, 18]. SC filters usually run at a fixed switching frequency in the slow-switching limit (SSL) range. In contrast, SC converters typically operate at variable frequency toward the boundary of the fast-switching limit (FSL) (see Sect. 2.3.2). In recent years, analytical methods of SC filter have been extended to SC converters.

In [10], the dynamic model is derived with multiple average equivalent sub-circuits similar to the derivation of the equivalent output resistance in Sect. 2.3.2. But its linearization requires a Spice-like simulator. A more systematic approach is introduced in [11–13] with a discrete-time dynamic model where the complex differential equations of the whole systems are solved in each phase. This method is very elaborate, while in most cases the internal dynamics of the converter are of limited interests. Sampled-data modeling [14, 15] is an alternative approach that investigates the behavior of the converter with the support of Spice-like simulators in order to avoid the complex solution of the state space equations.

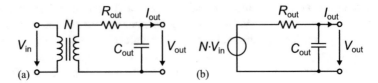

Fig. 6.18 (a) Dynamic SC converter model; (b) simplified dynamic model for calculation of the transfer function

Fig. 6.19 Transient response in SC operation mode to an input voltage step from 4 V to 3 V at an output current of $I_{out} = 20\,mA$

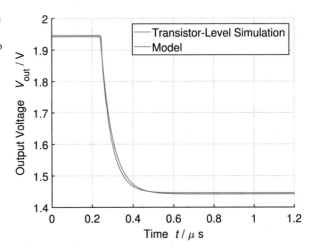

In the framework of this research, the static equivalent output resistance model is extended by the output capacitor C_{out}, as shown in Fig. 6.18a. This is a very useful approach and is also indicated in [15] where the output resistance R_{out} was derived with sampled-data modeling.

The transfer functions can be calculated from Fig. 6.18b where the DC transformer is substituted by an ideal voltage source $N \cdot V_{in}$.

$$\frac{V_{out}(s)}{V_{in}(s)} = \frac{N}{R_{out}C_{out}s + 1} \tag{6.3}$$

$$\frac{V_{out}(s)}{I_{out}(s)} = \frac{-R_{out}}{R_{out}C_{out}s + 1} \tag{6.4}$$

Figure 6.19 shows the transient response of this model and of a transistor-level simulation of the converter to an input voltage step from 4 V to 3 V. The model and the simulations are in a good agreement and can therefore be used for further investigations.

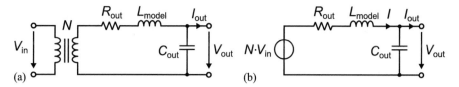

Fig. 6.20 (**a**) Dynamic ReSC converter model; (**b**) simplified dynamic model for calculation of the transfer function

6.2.2   Resonant SC Mode

For dynamic modeling of resonant SC converters, the introduced model in Sect. 6.2.1 for SC converters is extended by an inductor L_{model} as shown in Fig. 6.20a. In first order it corresponds to the actual inductance L (details below). For calculation of the transfer function, the DC transformer is substituted by an ideal voltage source $N \cdot V_{in}$, Fig. 6.20b.

Using Kirchhoff's voltage law, Fig. 6.20b leads to the following equation

$$N \cdot V_{in} = R_{out} \cdot I + L_{model} \cdot \frac{dI}{dt} + V_{out} \tag{6.5}$$

where the current I can be expressed by

$$V_{out} = \frac{1}{C_{out}} \int I - I_{out}\, dt \quad \Rightarrow \quad I = C_{out} \cdot \frac{dV_{out}}{dt} + I_{out}. \tag{6.6}$$

Inserting Eq. 6.6 into Eq. 6.5 leads to the following differential equation,

$$L_{model}C_{out} \cdot \frac{d^2 V_{out}}{dt^2} + R_{out}C_{out} \cdot \frac{dV_{out}}{dt} + V_{out} = N \cdot V_{in} - L_{model} \cdot \frac{dI_{out}}{dt} - R_{out} \cdot I_{out}, \tag{6.7}$$

which allows to derive the transfer functions

$$\frac{V_{out}(s)}{V_{in}(s)} = \frac{N}{L_{model}C_{out}s^2 + R_{out}C_{out}s + 1} \tag{6.8}$$

$$\frac{V_{out}(s)}{I_{out}(s)} = -\frac{L_{model}s + R_{out}}{L_{model}C_{out}s^2 + R_{out}C_{out}s + 1} \tag{6.9}$$

Figure 6.21a shows the line transient response of the model (Eq. 6.8) and a transistor-level simulation of the converter (similar to Fig. 6.19) to an input voltage step from 4 V to 3 V for $L_{model} = L$. A deviation between the model and the simulation is observed, in particular regarding the overshoots and undershoots. However, the static final values of the output voltage match, which again confirms the static equivalent output resistance model in Sect. 2.3.2. Furthermore, the slope

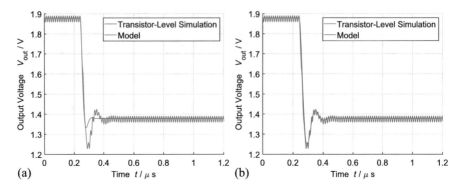

Fig. 6.21 Transient response in ReSC operation mode to an input voltage step from 4 V to 3 V at an output current of $I_{\text{out}} = 100\,\text{mA}$: (a) without correction factor, $L_{\text{model}} = L = 10\,\text{nH}$; (b) with correction factor, $L_{\text{model}} = K_{\text{corr}} \cdot L = 2.54 \cdot L$

at the beginning of the line step matches the simulation very well. This is mainly determined by the output capacitor C_{out}. For better matching, the inductance L_{model} has to be adjusted with an empirically determined correction factor K_{corr}, which can be used for all operation points. K_{corr} depends on the inductance value L of the converter, but it turned out to be independent of the converter's conversion ratio. For an inductance value of $L = 10\,\text{nH}$, a correction factor of $K_{\text{corr}} = 2.54$ was found. Figure 6.21b shows the transient response for that case. The model and the simulations are in a good agreement. Hence, the model can be used for further stability investigations.

The V_{out}-to-I_{out} transfer function (Eq. 6.9) is verified in Fig. 6.22 for a load step from 50 mA to 100 mA and $L_{\text{model}} = K_{\text{corr}} \cdot L$. It also confirms good matching between the model and the simulation results.

For the model of the different control loops, as will be investigated in Sect. 6.3, the plant transfer function $F_{\text{ReSC}}(s) = V_{\text{out}}(s) / R_{\text{out}}(s)$ is required. The differential equation (6.7) has to be linearized by a Taylor series expansion around the corresponding operation point ($I_{\text{out,OP}}$, $R_{\text{out,OP}}$, $V_{\text{out,OP}}$, $V_{\text{in,OP}}$) since it consists of multiple variables V_{out} and R_{out}. More details can be found in Appendix. Finally, the transfer function can be expressed by

$$\frac{V_{\text{out}}(s)}{R_{\text{out}}(s)} = F_{\text{ReSC}}(s) = \frac{-I_{\text{out,OP}}}{LC_{\text{out}}s^2 + R_{\text{out,OP}}C_{\text{out}}s + 1}. \tag{6.10}$$

The negative sign is compensated by the proportional term K_{Rout} introduced in Sect. 6.3.1. The linearization of the differential equation leads to a small error since the $R_{\text{out,OP}}$ in the dynamic component (see denominator polynomial in (6.10)) is not adjusted. However, this error is very small and can be neglected since the equivalent output resistance changes slowly and gradually (controlled by the counter). Furthermore, the worst case will be considered in the following analysis.

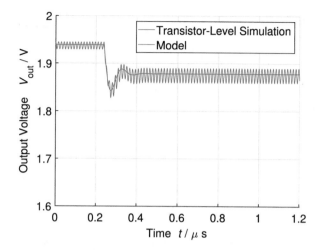

Fig. 6.22 Transient response in ReSC operation mode for a load step from 50 mA to 100 mA at an input voltage of $V_{in} = 4$ V and $L_{model} = K_{corr} \cdot L = 2.54 \cdot L$

6.3 Control Loop Model

Different control techniques for ReSC converters are introduced in Sect. 3.2, and their implementation is discussed in Sect. 6.1. In this section, the individual components of the SwCR, SC, and DOTM control loops are modeled analytically. Stability analyses are carried out for the different operation modes as well as for different conversion ratios N [19]. If the system is stable in all of these modes, it is assumed that it is also stable during a mode transition. First, the SwCR is investigated in Sect. 6.3.1, which is more complex due to its nonlinear behavior. No classical analysis tools such as Nyquist and Bode plots can be applied. After that, the frequency modulation in SC and DOTM mode is modeled in Sect. 6.3.2 with linear components.

The implemented model is also verified by experimental results which confirm its usability.

6.3.1 Model for Switch Conductance Regulation (SwCR)

In Sect. 3.2, the basic operation principle and, in Sect. 6.1, the implementation details of the switch conductance regulation (SwCR) were introduced. Figure 6.23 shows the implemented control loop together with a simplified block diagram. The control loop consists of the window comparator with a corresponding control logic, the counter, the segmented power switches, and the ReSC power stage (see also

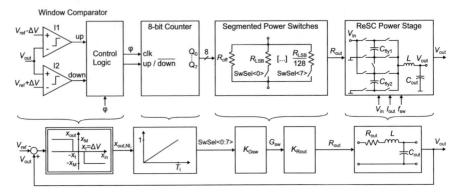

Fig. 6.23 Block diagram of the proposed SwCR control loop model

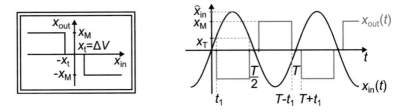

Fig. 6.24 Response of a relay with dead zone to a sinusoidal input

Sect. 6.1.1). The modeling approaches for the different blocks of the control loop are described below [19].

Modeling of the Window Comparator by Describing Function Analysis

The converter uses hysteretic control and tries to maintain the output voltage V_{out} within the hysteretic window ($V_{ref} - \Delta V$, $V_{ref} + \Delta V$), formed by a window comparator consisting of I1 and I2; see Fig. 6.23. The behavior of the window comparator is identical with an inverting nonlinear relay with dead zone as indicated in Fig. 6.24. Due to the nonlinear behavior, established stability analysis methods such as Nyquist and Bode plots cannot be directly applied. For analyzing such nonlinear time-invariant blocks, the describing function analysis, also called the method of harmonic balance, can be used [19–22]. The describing function analysis uses frequency domain (Fourier series) techniques to investigate limit cycle behavior in nonlinear systems. It can be considered as an extension of the Nyquist stability criterion to nonlinear systems.

Assume that a sinusoidal signal $x_{in} = \hat{x}_{in} \sin(\omega t)$ is applied to the input of a nonlinear relay with dead zone as shown in Fig. 6.24. The corresponding output x_{out} (also shown in Fig. 6.24) has the following values

$$x_{\text{out}}(t) = \begin{cases} 0 & -t_1 < t < t_1 \\ -x_M & t_1 < t < \frac{T}{2} - t_1 \\ 0 & \frac{T}{2} - t_1 < t < \frac{T}{2} + t_1 \\ x_M & \frac{T}{2} + t_1 < t < T - t_1 \end{cases} \tag{6.11}$$

The output x_{out} can be expressed by Fourier methods, which yields

$$x_{\text{out}} = \frac{a_0}{2} + \sum_{k=1}^{k=\infty} (a_k \cos(k\omega t) + b_k \sin(k\omega t)). \tag{6.12}$$

Figure 6.24 indicates that the output of the nonlinear relay has no DC component ($a_0 = 0$) but consists of a fundamental component ($a_1 \cos(k\omega t) + b_1 \sin(k\omega t)$) together with harmonic components at higher frequencies ($2\omega, 3\omega, \ldots$). The describing function method relies on neglecting all but the fundamental component $x_{\text{out},1}(t)$. Thereby it is assumed that the subsequent blocks have sufficient low-pass filter behavior, which is confirmed by subsequent tranistor-level simulations in Fig. 6.29. The nonlinear relay element can then be approximated by its describing function $N\left(\hat{x}_{\text{in}}\right)$ (also called equivalent gain) defined by the expression

$$N\left(\hat{x}_{\text{in}}\right) = \frac{x_{\text{out},1}(t)}{x_{\text{in}}(t)} = \frac{a_1 \cos(\omega t) + b_1 \sin(\omega t)}{\hat{x}_{\text{in}} \sin(\omega t)} = \frac{j a_1 \cdot e^{j\omega t} + b_1 \cdot e^{j\omega t}}{\hat{x}_{\text{in}} \cdot e^{j\omega t}} = \frac{j a_1 + b_1}{\hat{x}_{\text{in}}}. \tag{6.13}$$

Due to the odd symmetry of the output signal ($x_{\text{out}}(x_{\text{in}}) = -x_{\text{out}}(-x_{\text{in}})$), the Fourier coefficient a_1 is zero ($a_1 = 0$).

For the relay element with dead zone, only the Fourier coefficient b_1 has to be determined, which results in

$$b_1 = -\frac{4}{\pi} x_M \cdot \sqrt{1 - \left(\frac{x_t}{\hat{x}_{\text{in}}}\right)^2}. \tag{6.14}$$

The detailed calculations are shown in Appendix. Inserting (6.14) in (6.13), the describing function $N\left(\hat{x}_{\text{in}}\right)$ can be expressed by

$$N\left(\hat{x}_{\text{in}}\right) = -\frac{4}{\pi} \cdot \frac{x_M}{\hat{x}_{\text{in}}} \cdot \sqrt{1 - \left(\frac{x_t}{\hat{x}_{\text{in}}}\right)^2}. \tag{6.15}$$

Figure 6.25a shows the describing function $N\left(\hat{x}_{\text{in}}\right)$ versus the input signal $\hat{x}_{\text{in}}$. With increasing input signal $\hat{x}_{\text{in}}$, the describing function $N\left(\hat{x}_{\text{in}}\right)$ decreases until it reaches its minimum value $\hat{x}_{\text{in},x}$ which can be found by differentiation,

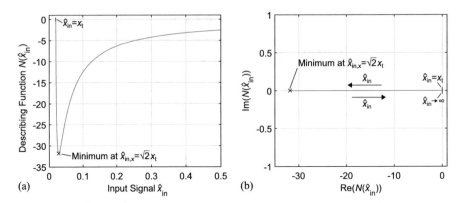

Fig. 6.25 (a) Describing function $N\left(\hat{x}_{\text{in}}\right)$ versus input signal x_{in} ($x_{\text{M}} = 1$, $x_{\text{t}} = 0.02$); (b) locus of the describing function $N\left(\hat{x}_{\text{in}}\right)$ in the complex plane

$$x_{\text{in,x}} = \sqrt{2} \cdot x_{\text{t}} \quad \Rightarrow \quad N(\hat{x}_{\text{in,x}}) = -\frac{2 \cdot x_{\text{M}}}{\pi \cdot x_{\text{t}}}. \tag{6.16}$$

Equation 6.16 is used in Sect. 6.4 for stability analysis. Since the amplitude of the output signal x_{out} of the relay is limited to x_{M}, the describing function $N\left(\hat{x}_{\text{in}}\right)$ approaches zero for $\hat{x}_{\text{in}} \rightarrow \infty$. Figure 6.25b shows the locus of the describing function $N\left(\hat{x}_{\text{in}}\right)$ in the complex plane. With increasing input signal $\hat{x}_{\text{in}}$, the locus $N\left(\hat{x}_{\text{in}}\right)$ moves from 0 along the negative real axis to the extremum (i.e., the minimum can be found in Eq. 6.16) and returns then to 0. Due to the purely real Fourier coefficient b_1, the describing function $N\left(\hat{x}_{\text{in}}\right)$ is also purely real. Thus, the relay block does not introduce any phase shift (see Fig. 6.25b).

Modeling of the Digital Counter

The 8-bit counter (see Fig. 6.23) can be modeled as a digital integrator with a integration time constant T_{I} together with a zero-order hold as shown in Fig. 6.26 [23–28]. As described in Sect. 6.1.1, the counter is clocked with the clock signal φ of the oscillator, which operates at the resonance frequency of the L-C_{fly} combination in the power stage. Clocking of the counter is modeled with the zero-order hold. For a general analysis of the control loop, $f_{\text{clk}} = 1/T_{\text{clk}}$ is used as the sampling frequency for the zero-order hold. The transfer function of the digital integrator can be derived from Fig. 6.26 as

$$F_{\text{DI}}(z) = \frac{1}{T_{\text{I}}} \cdot \frac{1}{1 - z^{-1}}. \tag{6.17}$$

For the system-level model shown in Fig. 6.23, it is beneficial to write the transfer function in the s-domain. With z^{-1} replaced by $e^{-sT_{\text{clk}}}$ and with consideration of

Fig. 6.26 Model of the counter

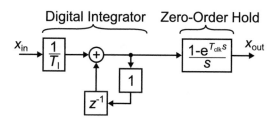

the zero-order hold, the transfer function of the digital counter can be written in the s-domain as

$$F_{counter}(s) = F_{DI}(s) \cdot F_H(s) = \frac{1}{T_I} \cdot \frac{1}{1 - e^{-sT_{clk}}} \cdot \frac{1 - e^{-sT_{clk}}}{s} = \frac{1}{s \cdot T_I}. \tag{6.18}$$

Therefore, the counter can be modeled as an ideal integrator block. Within a clock cycle, the counter value always changes by 1, independent from the amplitude of the input x_{in} which can be assumed to be constant for calculations. With these considerations, the time constant T_I can be calculated

$$\frac{1}{T_I} \cdot \int_0^{T_{clk}} x_{in} \, dt = 1 \quad \Rightarrow T_I = T_{clk} \cdot x_{in}. \tag{6.19}$$

The transfer function shows a dependency on the amplitude of the input signal x_{in}. However, this is always known since it is the output of the relay ($x_{in} = x_M$; see Fig. 6.23). It turns out that x_M gets eliminated during stability analysis (see Eq. 6.31).

Modeling of Segmented Power Switches and Equivalent Output Resistance

In order to model the segmented power switches, the 8-bit counter value SwSel<0:7> is converted into a conductance value G_{sw} and then into an equivalent output resistance R_{out} (see Fig. 6.23). Figure 3.10 and Eq. 3.9 in Sect. 3.2.2 show the linear relationship between the conversion of the counter value SwSel<0:7> and the conductance value G_{sw}. The offset in Eq. 3.9 can be omitted due to the small-signal modeling approach. This results in a simple proportional term K_{Gsw}

$$K_{Gsw} = G_{LSB}. \tag{6.20}$$

In the proposed switch implementation, this results in $K_{Gsw} = 7.2\,\text{mS}$ (see Sect. 3.2.2). Figure 6.27 shows the equivalent output resistance R_{out} versus the switch conductance value G_{sw} (Eq. 2.14). Due to the nonlinear behavior of the equivalent output resistance R_{out} to the switch conductance G_{sw} (see Sect. 4.2), a linearization at the operation point ($G_{sw,OP} = 1/R_{sw,OP}$) is necessary, which leads

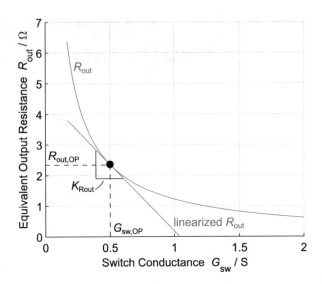

Fig. 6.27 Equivalent output resistance R_{out} versus switch conductance G_{sw} (conversion ratio $N = 1/2$)

to a proportional term K_{Rout}. Linearization is carried out for one LSB step G_{LSB} around the operation point $G_{sw,OP}$ as indicated in Fig. 6.27

$$K_{Rout} = \left.\frac{dR_{out}}{dG_{sw}}\right|_{OP} \approx \frac{R_{out}\left(G_{sw,OP} + G_{LSB}\right) - R_{out}\left(G_{sw,OP} - G_{LSB}\right)}{2 \cdot G_{LSB}} \qquad (6.21)$$

K_{Rout} strongly depends on the operation point which leads to a high gain at low counter values or low switch conductance G_{sw}.

Verification of the Model

In order to verify the model, the internal signals of the control loop are to be investigated. The feedback loop is opened and excited with a sinusoidal signal $x_{Excitation}$ with the frequency $f_{Excitation}$ as shown in Fig. 6.28. Then, the phase shifts between the internal signals are calculated with the model. These can be compared with the simulation results of a transistor-level simulation shown in Fig. 6.29. The inverting relay with dead time introduces a constant phase shift of $180°$ (see Fig. 6.24). The counter is modeled as an ideal integration element (see Eq. 6.18). This leads to a frequency-independent phase shift $-90°$, which corresponds to a delay of

$$\Delta t_{c,calc} = \frac{90°}{360°} \cdot T_{Excitation} = 140.9\,\text{ns}. \qquad (6.22)$$

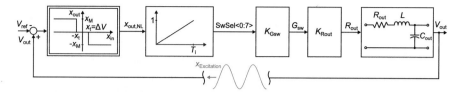

Fig. 6.28 Block diagram of the open control loop with sinusoidal excitation $x_{\text{Excitation}}$

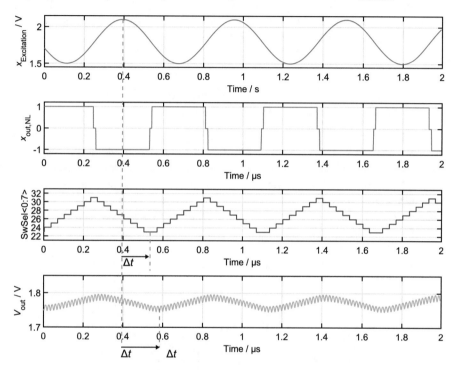

Fig. 6.29 Transistor-level simulation of the open control loop with a sinusoidal excitation

In Fig. 6.29, the simulated value of $\Delta t_{\text{c,sim}}$ is marked, starting from the maximum of the input signal $x_{\text{Excitation}}$ to the minimum of counter output SwSel<0:7> (first signal inversion of the relay is not included). The calculated value $\Delta t_{\text{c,calc}}$ fits well to the simulated value of $\Delta t_{\text{c,sim}} = 141$ ns. The proportional terms K_{Gsw} and K_{Rout} do not cause any phase shift. The second-order transfer function of the ReSC power stage introduces a frequency-dependent phase shift ϕ_{ReSC}, which can be derived from Eq. 6.10,

$$\phi_{\text{ReSC}} = \arg\left(\frac{I_{\text{out,OP}}}{LC_{\text{out}}(j\omega_{\text{Excitation}})^2 + R_{\text{out,OP}}C_{\text{out}}(j\omega_{\text{Excitation}}) + 1}\right). \quad (6.23)$$

This results in the delay of the power stage

$$\Delta t_{\text{ReSC,calc}} = \frac{\phi_{\text{ReSC}}}{360°} \cdot T_{\text{Excitation}} = 39 \text{ ns.} \qquad (6.24)$$

This calculated value also matches well with the $\Delta t_{\text{ReSC,sim}} = 41 \text{ ns}$ of the simulation in Fig. 6.29. Thus, a very good agreement between model and simulation is evident. The simulation also confirms that the assumption of a sufficient low-pass behavior of the linear blocks, required for harmonic balance analysis, is valid. The low-pass behavior of the counter and power stage is sufficient, so that only the first harmonic of the relay output $x_{\text{out,NL}}$ dominates the output signal V_{out}, which results in a sinusoidal waveform, as shown in Fig. 6.29.

6.3.2   Model for SC Control and Dynamic Off-Time Modulation (DOTM)

At low output power, the converter operates in SC mode in which the switching frequency is modulated by a hysteretic discrete-time control as described in Sect. 6.1.1. This controls the equivalent output resistance $R_{\text{out,SC}}$ (see Eq. 2.25) to maintain a constant output voltage. Exactly the same control concept is used for DOTM control where the equivalent output resistance $R_{\text{out,DOTM}}$ (see Eq. 2.28) is modulated by changing the time distance between the resonant pulses, as described in Sect. 6.1.2. In comparison to SwCR control, the frequency modulation of SC and DOTM control can be modeled much easier as depicted in Fig. 6.30.

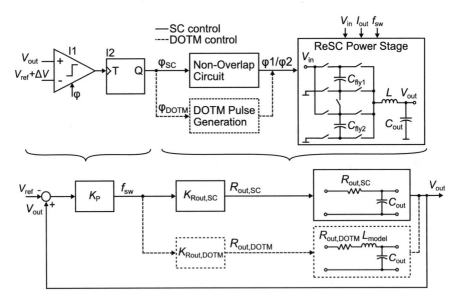

Fig. 6.30 Block diagram of the proposed modeling of the SC and DOTM control loop

The hysteretic discrete-time control consisting of the clocked comparator I1 and the T-flip-flop I2 can be combined to a simple proportional block K_P. The relation between the equivalent output resistance $R_{\text{out,SC}}$ or $R_{\text{out,DOTM}}$ and the switching frequency is described by Eq. 2.25 or 2.28, which has to be linearized at the respective operating point. This leads to the proportional factors $K_{\text{Rout,SC}}$ or $K_{\text{Rout,DOTM}}$. This is followed by the power stage of the resonant SC converter. As described in Sect. 6.2.1, the plant transfer function can be modeled as a first-order system in SC mode. For DOTM control, the second-order plant transfer function of the resonant SC converter has to be used, introduced in Sect. 6.5. Both transfer functions have to be linearized at the corresponding operating point, similar to Eq. 6.10 (see Appendix "Linearization of the Differential Equation of the Dynamic Model for the ReSC Converter").

Since the SC control loop in Fig. 6.30 is a first-order system, there is a maximum phase shift of $-90°$. Thus, in SC mode the converter is stable, because the phase margin is always $PM \geq 90°$. The DOTM control loop is a second-order system and is also stable at any operating point (see also experimental results in Sect. 6.1.6). This is mainly due to the low gain K_p of the hysteretic controller and the large external output capacitor C_{out} required in DOTM control, which leads to an overdamped system with a phase margin that is always $> 0°$. For this reason, the SC and DOTM mode will be no further analyzed regarding dynamic stability in Sect. 6.4.

6.4 Stability Analysis

With the model introduced in Sect. 6.3.1, stability analysis can be performed, and critical operation points can be identified. However, standard stability measures like Routh-Hurwitz criterion, Barkhausen criterion, Nyquist criterion, or root locus analysis cannot be used since the SwCR control loop contains nonlinear components [20]. This work proposes to apply the dual locus method [19–22], which is similar to the Nyquist procedure. It is an approximation method that can be applied for analyzing the existence and properties of limit cycle oscillations in nonlinear systems. Limit cycles are periodic oscillations performed by nonlinear systems. Without external input signals, the oscillations sustain with a certain frequency and amplitude. Since the describing function $N\left(\hat{x}_{\text{in}}\right)$ of the window comparators (see Eq. 6.15) cannot be combined with the other transfer functions (see Eqs. 6.10, 6.18, 6.20, and 6.21) in the s-plane, the two locus curves are plotted separately into the complex plane.

The dual locus method assumes that the open-loop system is a series connection of a single nonlinearity $N\left(\hat{x}_{\text{in}}\right)$ and a remaining linear part $F_{\text{linears}}(s)$ as depicted in Fig. 6.31. All linear transfer functions of the SwCR control loop (see Sect. 6.3.1)

Fig. 6.31 Splitting of the SwCR control loop into linear and nonlinear components

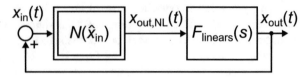

are combined in

$$F_{\text{linears}}(s) = F_{\text{counter}}(s) \cdot K_{\text{Gsw}} \cdot K_{\text{Rout}} \cdot F_{\text{ReSC}}(s)$$

$$= \frac{1}{x_M \cdot T_{\text{clk}}} \cdot \frac{1}{s} \cdot K_{\text{Gsw}} \cdot K_{\text{Rout}} \frac{-I_{\text{out,OP}}}{L C_{\text{out}} s^2 + R_{\text{out,OP}} C_{\text{out}} s + 1}, \qquad (6.25)$$

while the nonlinear transfer function is represented by the describing function $N\left(\hat{x}_{\text{in}}\right)$ of the window comparators (see Eq. 6.15), repeated here for convenience

$$N\left(\hat{x}_{\text{in}}\right) = -\frac{4}{\pi} \cdot \frac{x_M}{\hat{x}_{\text{in}}} \cdot \sqrt{1 - \left(\frac{x_t}{\hat{x}_{\text{in}}}\right)^2}. \qquad (6.26)$$

In Fig. 6.31, there is a positive feedback of the output signal, since the negative sign was taken into account in the describing function $N\left(\hat{x}_{\text{in}}\right)$ of the window comparators. The analysis of limit cycles starts with the assumption that the system is in the state of a sustained oscillation, described by

$$x_{\text{in}}(t) = x_{\text{out}}(t) = \hat{x}_{\text{out}} \cdot \sin\left(\omega \cdot t\right). \qquad (6.27)$$

A permanent oscillation in the loop appears for (see Fig. 6.31)

$$x_{\text{out}}(t) = \hat{x}_{\text{out}} \cdot \underbrace{\left|N\left(\hat{x}_{\text{in}}\right)\right| \cdot \left|F_{\text{linears}}\left(j\omega\right)\right|}_{\overset{!}{=}1} \cdot \sin\left(\omega t + \underbrace{\arg\{N\left(\hat{x}_{\text{in}}\right)\} + \arg\{F_{\text{linears}}\left(j\omega\right)\}}_{\overset{!}{=}-n\cdot360°}\right).$$

$$\qquad (6.28)$$

This can be described by the equation

$$N\left(\hat{x}_{\text{in}}\right) \cdot F_{\text{linears}}\left(j\omega\right) = 1 \qquad (6.29)$$

or

$$F_{\text{linears}}\left(j\omega\right) = \frac{1}{N\left(\hat{x}_{\text{in}}\right)}. \qquad (6.30)$$

Equation 6.30 is similar to the Nyquist criterion for the linear case ($F_{\text{linears}}\left(j\omega\right) = -1$). There, the locus $F_{\text{linears}}\left(j\omega\right)$ is examined with respect to the critical point -1, while here the locus of $F_{\text{linears}}\left(j\omega\right)$ is examined with respect to the inverse describing function $1/N\left(\hat{x}_{\text{in}}\right)$. Both loci F_{linears} and $N\left(\hat{x}_{\text{in}}\right)$ can be plotted in the

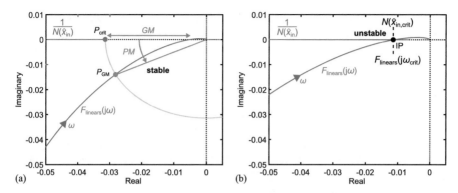

Fig. 6.32 Dual locus plot of the inverse describing function $1/N\left(\hat{x}_{in}\right)$ and the linear transfer function $F_{linears}\left(j\omega\right)$ for a stable **(a)** and an unstable system **(b)**

complex plane since they are respective functions of ω and $\hat{x}_{in}$ only. The system in Fig. 6.32a is stable, since there is no intersection between the locus $F_{linears}\left(j\omega\right)$ of the linear part and the locus of the inverse describing function $1/N\left(\hat{x}_{in}\right)$. A critical point P_{crit} can be identified from the locus of the inverse describing function $1/N\left(\hat{x}_{in}\right)$ as marked in Fig. 6.32a. If this point is exceeded by the locus $F_{linears}\left(j\omega\right)$, the system becomes unstable. The critical point P_{crit} can also be used to determine the gain margin GM and the phase margin PM. The gain margin GM is the quotient between the critical point P_{crit} and the intersection of the $F_{linears}\left(j\omega\right)$ locus with the real axis. The phase margin PM is the angle to the real axis where the magnitude of $F_{linears}\left(j\omega\right)$ equals the critical point P_{crit}. In addition, a circle can be inserted whose radius exactly corresponds to the absolute value of the critical point (similar to the unit circle in the Nyquist criterion). In that case, the phase margin PM can be determined from the intersection P_{GM} of the circle and $F_{linears}\left(j\omega\right)$ with $PM = \arctan\left(\text{Im}\left(P_{GM}\right)/\text{Re}\left(P_{GM}\right)\right)$.

Figure 6.32b shows an unstable operation point. Using the values $N(\hat{x}_{in,crit})$ and $F_{linears}(\omega_{crit})$ of both loci at the intersection point IP, the amplitude $\hat{x}_{in,crit}$ and frequency ω_{crit} of the limit cycle oscillation can be calculated.

In the following, a useful stability criterion for the design of the control loop is derived. Equations 6.25, 6.26, and 6.30 can be used to perform further calculations (see Appendix), which lead to the following equation

$$\frac{K_{Gsw} \cdot |K_{Rout}| \cdot I_{out,OP} \cdot L}{T_{clk} \cdot R_{out,OP}} \leq \frac{\pi \cdot x_t}{2}. \tag{6.31}$$

Equation 6.31 is a very simple and easy-to-use criterion for designing the control and the converter regarding stability and robustness. In the following, it will be examined, which design and operating parameters can be specifically influenced to ensure the stability of the converter.

The inductance L, as well as the output capacitor C_{out}, the flying capacitors C_{fly}, the power switches (K_{Gsw}, K_{Rout}), and the equivalent output resistance R_{out}, cannot be chosen arbitrarily. They either are determined by the application or need to be designed based on the efficiency analyses (see Sect. 3.3). The hysteretic window of the window comparator (i.e., the dead zone $x_t = \Delta V$ of the relay) is chosen according to the specification of the output voltage ripple. The output voltage ripple should be smaller than as the hysteretic window. However, a larger window also increases stability according to Eq. 6.31. In the proposed converter (see Sect. 6.1.1), a window size of $x_t = 20\,\text{mV}$ is chosen. In conclusion, the clock frequency $f_{clk} = 1/T_{clk}$ of the counter is the only freely adjustable parameter in Eq. 6.31 that affects the stability of the control loop. It directly affects the integration time constant T_I and therefore the gain of the 8-bit counter (see Eqs. 6.18 and 6.19).

The impact of different clock frequencies f_{clk} and window sizes x_t on the dual locus plot is shown in Fig. 6.33. The critical point P_{crit} is determined by the absolute maximum of the describing function $N(\hat{x}_{in})$ and can be calculated from Eq. 6.16 to be $P_{crit} = -(\pi \cdot x_t)/2$. With a smaller window size x_t, the critical point P_{crit} moves closer to the locus of $F_{linears}(j\omega)$. A small comparator window is therefore more likely to cause instability. With increased clock frequency, the $F_{linears}(j\omega)$ locus approaches the real axis, resulting in a smaller phase margin PM. With a small window size of $x_t = 5\,\text{mV}$ and high clock frequencies of $3 f_{sw,res}$ or $4 f_{sw,res}$, instabilities can be provoked, which can be seen at the intersection of the inverse describing function locus $1/N(\hat{x}_{in})$ with the $F_{linears}(j\omega)$ locus in Fig. 6.33. An operating point with $f_{clk} = 3.4 f_{sw,res}$ is also experimentally verified in Sect. 6.5.

Since there is a simple equation for evaluation of the stability (see Eq. 6.31), it is advantageous to consider the worst-case scenario in the following. For the implemented control loop (see Sect. 6.1.1), a window size of $x_t = 20\,\text{mV}$ and a clock frequency of $f_{clk} = f_{sw,res}$ are chosen. The clock frequency varies depending on the conversion ratio N ($f_{sw,res} = 35.5\,\text{MHz}$ for $N = 1/2$ or $f_{sw,res} = 47\,\text{MHz}$ for $N = x/3$). According to Eq. 6.31, the worst case for the stability occurs at a counter value SwSel<0:7>= 0 ($R_{sw,max}$) since this corresponds to the highest gain $|K_{Rout}|$. Due to the steepness of the hyperbola at low switch conductance values or low counter values (Fig. 3.10), $|K_{Rout}|$ is particularly high at this point and outweighs the equivalent output value $R_{out,OP}$. In addition, the highest possible current $I_{out,OP}$ has to be chosen for the worst case. Table 6.2 shows the resulting worst-case operating points for this converter design. The $R_{out,OP,wc}$ values (calculated with Eq. 2.14) result from the maximum switch resistance $R_{sw,max}$. For $V_{in,wc}$, the maximum input voltage of the respective ratio was chosen, because there, the highest output currents for large resistance values (required for regulation) result. $I_{out,OP,wc}$ can then be calculated according to the equivalent output resistance model with Eq. 3.4. Figure 6.34 shows the dual locus plot for these operating points. There is no intersection of the inverse describing function locus $1/N(\hat{x}_{in})$ with the $F_{linears}(j\omega)$ loci. This means that the converter is stable for all operating points within the switch conductance regulation (SwCR). In ratio 1/2, a phase margin

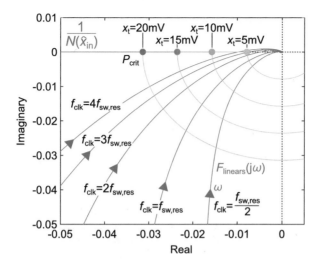

Fig. 6.33 Dual locus plot of the inverse describing function $1/N\left(\hat{x}_{in}\right)$ and the linear transfer function $F_{linears}$ for different clock frequencies f_{clk} and different window sizes x_t for the dead zone ($N = 1/2$, $V_{in} = 4.5\,\text{V}$, $I_{out,OP} = 72\,\text{mA}$, $V_{out} = 1.8\,\text{V}$)

Table 6.2 Worse-case operating points

Parameter	$N=1/2$	$N=2/3$
$R_{out,OP,wc}$	$6.4\,\Omega$	$10.9\,\Omega$
$I_{out,OP,wc}$	$72.6\,\text{mA}$	$60.7\,\text{mA}$
$V_{in,wc}$	$4.5\,\text{V}$	$3.7\,\text{V}$

of $PM = 48°$ is reached, while in ratio 2/3, a phase margin of $PM = 37°$ is obtained. Since this is the worst-case scenario, this is sufficient for the evaluation of the converter stability.

6.5 Experimental Results

This section examines if the model is able to predict the actual behavior of the ReSC converter. As proven by simulations and measurement results in Sect. 6.1.6, the proposed converter is stable at all operating points. In order to verify the stability model with measurement results of the ReSC converter, a stable and an unstable operating point is investigated for this purpose. As indicated in Sect. 6.4, instabilities can be created by significantly increasing the clock frequency f_{clk} and by decreasing the comparator window x_t. An unstable operating point can be created with a clock frequency $f_{clk} = 3.4 \cdot f_{sw,res} = 120\,\text{MHz}$ (see Fig. 6.33) and a comparator window of $x_t = 5\,\text{mV}$ at an input voltage of $V_{in} = 4.3\,\text{V}$ and an output current of $I_{out} = 40\,\text{mA}$. With these parameters, the dual locus plot in Fig. 6.35a shows an

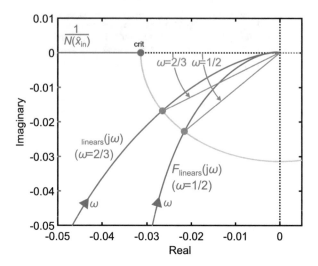

Fig. 6.34 Dual locus plot of the inverse describing function $1/N\left(\hat{x}_{in}\right)$ and the linear transfer function F_{linears} for the worst-case operating points

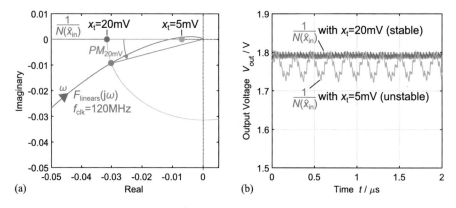

Fig. 6.35 Validation of the proposed model with measurement results. Intentional creation of instabilities with increasing clock frequency f_{clk} and smaller comparator window x_t: **(a)** dual locus plot for a stable ($x_t = 20\,\text{mV}$) and an unstable ($x_t = 5\,\text{mV}$) operating point at a clock frequency of $f_{\text{clk}}=120\,\text{MHz}$; **(b)** transient measurements of the output voltage V_{out} for a stable and unstable operating point predicted by the proposed model

intersection of the inverse describing function locus $1/N\left(\hat{x}_{in}\right)$ with the $F_{\text{linears}}\left(j\omega\right)$ locus predicting instability. This is confirmed by transient measurements of the output voltage V_{out} in Fig. 6.35b, which shows oscillations for the output voltage. With a larger comparator window $x_t = 20\,\text{mV}$, the dual locus plot predicts a stable operation with a phase margin of $PM_{20\text{mV}} = 17°$. A stable operation can also be observed in the transient measurement results in Fig. 6.35b.

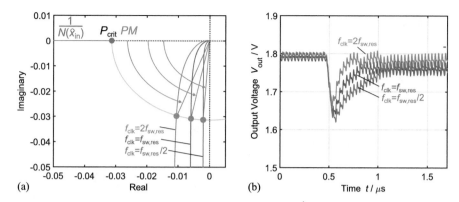

Fig. 6.36 Validation of the proposed model with measurement results. Investigation of the influence of the clock frequency on the phase margin: (**a**) dual locus plot for different clock frequencies f_{clk}; (**b**) measured transient load step response for different clock frequencies f_{clk}

In order to investigate the influence of the clock frequency of the counter f_{clk} on the phase margin and response time, the dual locus plot and the load step response of the converter for different clock frequencies are shown in Fig. 6.36. The clock frequency directly determines the gain of the control loop (see Eq. 6.18), which then regulates faster or slower during load steps. Thus, the behavior of the converter can be evaluated at different integration time constants. A load step of 30 mA to 120 mA is performed at an input voltage of $V_{\text{in}} = 3.9\,\text{V}$. As expected, the lowest clock frequency $f_{\text{clk}} = f_{\text{sw,res}}/2 = 17.75\,\text{MHz}$ leads to a high phase margin of $PM = 85.3°$ (Fig. 6.36a). Due to the small loop gain, however, this results in the longest settling time of 500 ns (Fig. 6.36b). A clock frequency of $f_{\text{clk}} = 2f_{\text{sw,res}} = 35.5\,\text{MHz}$ results in a phase margin of $PM = 78.3°$ and a response time of 340 ns. The lowest phase margin of $PM = 69.2°$ occurs at the highest clock frequency of $f_{\text{clk}} = 2f_{\text{sw,res}} = 71\,\text{MHz}$. There, the fastest response time of 200 ns is achieved. Overall, theory and experimental results match very well. Hence, these results confirm the proposed stability model and in particular the stability condition of Eq. 6.31.

Appendix

Linearization of the Differential Equation of the Dynamic Model for the ReSC Converter

The following differential equation (6.32) has to be linearized by a Taylor series expansion around the corresponding operation point ($I_{out,OP}$, $R_{out,OP}$, $V_{out,OP}$, $V_{in,OP}$) since it consists of multiple variables V_{out} and R_{out}.

$$L_{model}C_{out} \cdot \frac{d^2 V_{out}}{dt^2} + R_{out}C_{out} \cdot \frac{dV_{out}}{dt} + V_{out} = N \cdot V_{in} - L_{model} \cdot \frac{dI_{out}}{dt} - R_{out} \cdot I_{out},$$
(6.32)

For the definition of the operating point, the input voltage V_{in} and the output current I_{out} are assumed to be constant

$$I_{out,OP} = I_{out} = \text{const} \quad \Rightarrow \quad \ddot{I}_{out,OP} = \dot{I}_{out,OP} = 0; \quad i_{out} = 0$$

$$V_{in,OP} = V_{out} = \text{const} \quad \Rightarrow \quad \ddot{V}_{in,OP} = \dot{V}_{in,OP} = 0; \quad v_{in} = 0$$

$$V_{out,OP} = \text{const} \quad \Rightarrow \quad \ddot{V}_{out,OP} = \dot{V}_{out,OP} = 0$$

$$R_{out,OP} = \text{const} \quad \Rightarrow \quad \ddot{R}_{out,OP} = \dot{R}_{out,OP} = 0$$

With this, Eq. 6.32 can be rewritten (dot notation is used for differential operator for simplicity)

$$f(\ddot{V}_{out}, \dot{V}_{out}, V_{out}, R_{out}) = L_{model}C_{out} \cdot \ddot{V}_{out} + R_{out}C_{out} \cdot \dot{V}_{out} + V_{out} - N \cdot V_{in,OP}$$
$$+ L_{model} \cdot \dot{I}_{out,OP} + R_{out} \cdot I_{out,OP} = 0.$$
(6.33)

The linearized differential equation can be derived by

$$\frac{\partial f}{\partial \ddot{V}_{out}}\Big|_{OP} \cdot \ddot{v}_{out} + \frac{\partial f}{\partial \dot{V}_{out}}\Big|_{OP} \cdot \dot{v}_{out} + \frac{\partial f}{\partial V_{out}}\Big|_{OP} \cdot v_{out} + \frac{\partial f}{\partial R_{out}}\Big|_{OP} \cdot r_{out} = 0,$$
(6.34)

where v_{out} and r_{out} are the small-signal changes around the operation point (e.g., ($v_{out} = V_{out} - V_{out,OP}$)). Solving Eq. 6.34 leads to

$$L_{model}C_{out} \cdot \ddot{v}_{out} + R_{out,OP}C_{out} \cdot \dot{v}_{out} + v_{out} = -I_{out,OP} \cdot r_{out}.$$
(6.35)

Now, the plant transfer function for the small signals (e.g., $v_{out} = V_{out} - V_{out,OP}$) can be derived from Eq. 6.35

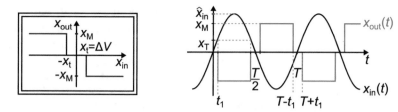

Fig. 6.37 Response of a relay with dead zone to a sinusoidal input

$$\frac{v_{out}(s)}{r_{out}(s)} = F_{ReSC}(s) = \frac{-I_{out,OP}}{LC_{out}s^2 + R_{out,OP}C_{out}s + 1}. \tag{6.36}$$

Window Comparator Model Based on the Describing Function Analysis

The describing function analysis (also called the method of harmonic balance) uses frequency domain (Fourier series) techniques to investigate limit cycle behavior in nonlinear systems. Assume that a sinusoidal signal $x_{in} = \hat{x}_{in} \sin(\omega t)$ is input to the nonlinear relay with dead zone as shown in Fig. 6.37. With this, the corresponding output x_{out} (Fig. 6.24) has the following values

$$x_{out}(t) = \begin{cases} 0 & -t_1 < t < t_1 \\ -x_M & t_1 < t < \frac{T}{2} - t_1 \\ 0 & \frac{T}{2} - t_1 < t < \frac{T}{2} + t_1 \\ x_M & \frac{T}{2} + t_1 < t < T - t_1. \end{cases} \tag{6.37}$$

The input x_{in} reached the threshold x_t of the dead zone at $t = t_1$, which can be calculated with

$$x_{in}(t = t_1) = x_t = \hat{x}_{in} \cdot \sin\left(\frac{2\pi}{T} \cdot t_1\right)$$

$$\Rightarrow t_1 = \frac{T}{2\pi} \cdot \arcsin\left(\frac{x_t}{\hat{x}_{in}}\right) \tag{6.38}$$

The output x_{out} can be expressed by Fourier methods with

$$x_{out} = \frac{a_0}{2} + \sum_{k=1}^{k=\infty} (a_k \cos(k\omega t) + b_k \sin(k\omega t)). \tag{6.39}$$

The output of the nonlinear relay has no DC component ($a_0 = 0$) but consists of a fundamental component ($a_1 \cos(k\omega t) + b_1 \sin(k\omega t)$) together with harmonic components at higher frequencies ($2\omega, 3\omega, \ldots$). The describing function method relies on neglecting all but the fundamental component $x_{\text{out},1}(t)$. Thereby, it is assumed that the subsequent blocks have sufficient low-pass filter behavior. The nonlinear relay element can then be approximated by its describing function $N\left(\hat{x}_{\text{in}}\right)$ (also called equivalent gain) defined by the expression

$$N\left(\hat{x}_{\text{in}}\right) = \frac{x_{\text{out},1}(t)}{x_{\text{in}}(t)} = \frac{a_1 \cos(\omega t) + b_1 \sin(\omega t)}{\hat{x}_{\text{in}} \sin(\omega t)} = \frac{j a_1 \cdot e^{j\omega t} + b_1 \cdot e^{j\omega t}}{den} = \frac{j a_1 + b_1}{\hat{x}_{\text{in}}}.$$

$$(6.40)$$

The Fourier coefficients a_1 and b_1 can be calculated with

$$a_1 = \frac{2}{T} \int_0^T x_{\text{out}}(t) \cdot \cos(\omega t) dt \qquad (6.41)$$

$$b_1 = \frac{2}{T} \int_0^T x_{\text{out}}(t) \cdot \sin(\omega t) dt. \qquad (6.42)$$

Due to the odd symmetry of the output signal ($x_{\text{out}}(x_{\text{in}}) = -x_{\text{out}}(-x_{\text{in}})$), the Fourier coefficients a_k are zero ($a_1 = 0$). In this case, the Fourier coefficient b_1 can be calculated with

$$b_1 = \frac{4}{T} \int_0^{T/2} x_{\text{out}}(t) \cdot \sin(\omega t) dt \qquad (6.43)$$

Due to the symmetry of x_{out} to $t = T/4$, the calculation of b_1 can be simplified to

$$b_1 = \frac{8}{T} \int_{t_1}^{T/4} x_{\text{out}}(t) \cdot \sin(\omega t) \, dt = \frac{8}{T} \int_{t_1}^{T/4} -x_{\text{M}} \cdot \sin(\omega t) dt \qquad (6.44)$$

Solving of Eq. 6.44 leads to

$$b_1 = -\frac{8}{T} \frac{x_{\text{M}}}{\omega} \left[-\cos(\omega t) \right]_{t_1}^{T/4} = -\frac{4}{\pi} \cdot x_{\text{M}} \left[-\cos\left(\frac{2\pi}{T} \cdot t\right) \right]_{t_1}^{T/4}$$

$$= -\frac{4}{\pi} \cdot x_{\text{M}} \left(\underbrace{-\cos\left(\frac{\pi}{2}\right)}_{=0} + \cos\left(\frac{2\pi}{T} \cdot t_1\right) \right) \qquad (6.45)$$

$$= -\frac{4}{\pi} \cdot x_{\text{M}} \left(\cos\left(\frac{2\pi}{T} \cdot t_1\right) \right)$$

Replacing t_1 (Eq. 6.38) with Eq. 6.44 leads to

$$b_1 = -\frac{4}{\pi} \cdot x_M \left(\cos \left(\arcsin \left(\frac{x_t}{\hat{x}_{in}} \right) \right) \right) \tag{6.46}$$

With $\cos(\arcsin(x)) = \sqrt{1 - x^2}$, b_1 can be expressed as

$$b_1 = -\frac{4}{\pi} x_M \cdot \sqrt{1 - \left(\frac{x_t}{\hat{x}_{in}} \right)^2}. \tag{6.47}$$

Now, the describing function $N\left(\hat{x}_{in}\right)$ can be expressed with

$$N\left(\hat{x}_{in}\right) = -\frac{4}{\pi} \cdot \frac{x_M}{\hat{x}_{in}} \cdot \sqrt{1 - \left(\frac{x_t}{\hat{x}_{in}} \right)^2}. \tag{6.48}$$

Extension of the Dual Locus Method by Analytical Investigations

To investigate the influence of various parameters on the stability of the SwCR control loop of the ReSC converter, Eqs. 6.49, 6.50, and 6.51 can be used to perform further calculations.

$$F_{\text{linears}}\left(j\omega\right) = \frac{1}{N\left(\hat{x}_{in}\right)} \tag{6.49}$$

$$N\left(\hat{x}_{in}\right) = -\frac{4}{\pi} \cdot \frac{x_M}{\hat{x}_{in}} \cdot \sqrt{1 - \left(\frac{x_t}{\hat{x}_{in}} \right)^2}. \tag{6.50}$$

$$F_{\text{linears}}\left(j\omega\right) = F_{\text{counter}}\left(j\omega\right) \cdot K_{\text{Gsw}} \cdot K_{\text{Rout}} \cdot F_{\text{ReSC}}\left(j\omega\right)$$

$$= \frac{1}{x_M \cdot T_{\text{clk}}} \cdot \frac{1}{(j\omega)} \cdot K_{\text{Gsw}} \cdot K_{\text{Rout}} \frac{-I_{\text{out,OP}}}{L C_{\text{out}}\left(j\omega\right)^2 + R_{\text{out,OP}} C_{\text{out}}\left(j\omega\right) + 1} \tag{6.51}$$

First, the phase condition is used to calculate the critical frequency ω_{crit}, which leads to an instability. Since the describing function $N\left(\hat{x}_{in}\right)$ is purely real and has a negative sign, a constant phase shift of $\arg\left(N\left(\hat{x}_{in}\right)\right) = -180°$ is obtained.

$$\arg\left(N\left(\hat{x}_{\text{in}}\right)\right) + \arg\left(F_{\text{linears}}\left(j\omega_{\text{crit}}\right)\right) = -180° + \arg\left(F_{\text{linears}}\left(j\omega_{\text{crit}}\right)\right) = -n \cdot 360° \tag{6.52}$$

With $n = 1$, a phase shift of $-180°$ is missing to fulfill the phase condition.

$$\Rightarrow \arg\left(F_{\text{linears}}\left(j\omega_{\text{crit}}\right)\right) = -180° \tag{6.53}$$

Thereby, a phase shift of $-90°$ comes from the integrator block (8-bit counter). The remaining phase shift of $-90°$ comes from the second-order transfer function of the ReSC power stage which occurs at the natural frequency ω_o of the converter.

$$\Rightarrow \omega_{\text{crit}} = \omega_o = \frac{1}{\sqrt{LC_{\text{out}}}} \tag{6.54}$$

The critical frequency ω_{crit} can now be used to evaluate the amplitude condition

$$\left|F_{\text{linears}}\left(j\omega_{\text{crit}}\right)\right| = \left|\frac{K_{\text{Gsw}} \cdot K_{\text{Rout}} \cdot I_{\text{out,OP}} \cdot L}{x_M \cdot T_{\text{clk}} \cdot R_{\text{out,OP}}}\right| \leq \frac{1}{\left|N\left(\hat{x}_{\text{in}}\right)\right|} \tag{6.55}$$

The signs of the variables K_{Gsw}, $I_{\text{out,OP}}$, T_{clk}, $R_{\text{out,OP}}$, and x_M are always positive, while the sign of K_{Rout} is always negative.

$$-\frac{K_{\text{Gsw}} \cdot |K_{\text{Rout}}| \cdot I_{\text{out,OP}} \cdot L}{x_M \cdot T_{\text{clk}} \cdot R_{\text{out,OP}}} \leq \frac{1}{\left|N\left(\hat{x}_{\text{in}}\right)\right|} \tag{6.56}$$

The minimum of the describing function $N\left(\hat{x}_{\text{in}}\right)$ can be calculated with

$$N(\hat{x}_{\text{in,x}}) = -\frac{2 \cdot x_M}{\pi \cdot x_t}. \tag{6.57}$$

This can be inserted in Eq. 6.56

$$\frac{K_{\text{Gsw}} \cdot |K_{\text{Rout}}| \cdot I_{\text{out,OP}} \cdot L}{x_M \cdot T_{\text{clk}} \cdot R_{\text{out,OP}}} \leq \frac{\pi \cdot x_t}{2}. \tag{6.58}$$

With Eq. 6.58, the influence of various parameters on the stability of the SwCR control loop can be investigated. Therefore it can be used for design of the control loop.

References

1. Baek, J., et al.: Switched inductor capacitor buck converter with >85% power efficiency in 100uA-to-300mA loads using a bang-bang zero-current detector. In: 2018 IEEE Custom Integrated Circuits Conference (CICC), pp. 1–4 (2018). https://doi.org/10.1109/CICC.2018.8357024
2. El-Chammas, M., Murmann, B.: A 12-GS/s 81-mW 5-bit time-interleaved flash ADC with background timing skew calibration. IEEE J. Solid-State Circuits 46(4), 838–847 (2011). ISSN: 1558-173X. https://doi.org/10.1109/JSSC.2011.2108125
3. Park, S., Palaskas, Y., Flynn, M.P.: A 4-GS/s 4-bit flash ADC in 0.18- μm CMOS. IEEE J. Solid-State Circuits 42(9), 1865–1872 (2007). ISSN: 1558-173X. https://doi.org/10.1109/JSSC.2007.903053
4. Wicht, B.: Current Sense Amplifiers for Embedded SRAM in High-Performance System-on-a-Chip Designs. Springer Series in Advanced Micro-electronics. Springer, Berlin/Heidelberg (2013). ISBN: 978-3-540-00298-7. https://books.google.de/books?id=E-7tCAAAQBAJ
5. Kobayashi, T., et al.: A current-mode latch sense amplifier and a static power saving input buffer for low-power architecture. In: 1992 Symposium on VLSI Circuits Digest of Technical Papers, pp. 28–29 (1992). https://doi.org/10.1109/VLSIC.1992.229252
6. Wicht, B., Nirschl, T., Schmitt-Landsiedel, D.: Yield and speed optimization of a latch-type voltage sense amplifier. IEEE J. Solid-State Circuits 39(7), 1148–1158 (2004)
7. Baker, R.J.: CMOS Circuit Design, Layout, and Simulation, 3rd edn. Wiley-IEEE Press (2010). ISBN: 0470881321
8. Schinkel, D., et al.: (2007). A double-tail latch-type voltage sense amplifier with 18ps setup+hold time. In: 2007 IEEE International Solid-State Circuits Conference. Digest of Technical Papers, pp. 314–605. https://doi.org/10.1109/ISSCC.2007.373420
9. Allen, P.E., Holberg, D.R.: CMOS Analog Circuit Design. Oxford University Press, New York (2012)
10. Ben-Yaakov, S., Evzelman, M.: Generic average modeling and simulation of the static and dynamic behavior of switched capacitor converters. In: 2012 Twenty-Seventh Annual IEEE Applied Power Electronics Conference and Exposition (APEC), pp. 2568–2575 (2012). https://doi.org/10.1109/APEC.2012.6166185
11. Henry, J.M., Kimball, J.W.: Practical performance analysis of complex switched-capacitor converters. IEEE Trans. Power Electron. 26(1), 127–136 (2011). ISSN: 1941-0107. https://doi.org/10.1109/TPEL.2010.2052634
12. Mayo-Maldonado, J.C., Rosas-Caro, J.C., Rapisarda, P.: Modeling approaches for DC–DC converters with switched capacitors. IEEE Trans. Ind. Electron. 62(2), 953–959 (2015). ISSN: 1557-9948. https://doi.org/10.1109/TIE.2014.2353013
13. Müller, L., Kimball, J.W.: A dynamic model of switched-capacitor power converters. IEEE Trans. Power Electron. 29(4), 1862–1869 (2014). ISSN: 1941-0107. https://doi.org/10.1109/TPEL.2013.2264756
14. Souvignet, T., Allard, B., Lin-Shi, X.: Sampled-data modeling of switched-capacitor voltage regulator with frequency-modulation control. IEEE Trans. Circuits Syst. I Regul. Pap. 62(4), 957–966 (2015). ISSN: 1558-0806. https://doi.org/10.1109/TCSI.2015.2399025
15. Souvignet, T., Allard, B., Trochut, S.: A fully integrated switched-capacitor regulator with frequency modulation control in 28-nm FDSOI. IEEE Trans. Power Electron. 31(7), 4984–4994 (2016). ISSN: 1941-0107. https://doi.org/10.1109/TPEL.2015.2478850
16. Seeman, M.D.: A Design Methodology for Switched-Capacitor DCDC Converters. Ph.D. thesis. EECS Department, University of California, Berkeley (2009). http://www2.eecs.berkeley.edu/Pubs/TechRpts/2009/EECS-2009-78.html
17. Tsividis, Y.: Analysis of switched capacitive networks. IEEE Trans. Circuits Syst. 26(11), 935–947 (1979). ISSN: 1558-1276. https://doi.org/10.1109/TCS.1979.1084588

18. Fang, S-C., Tsividis, Y., Wing, O.: Time- and frequency domain analysis of linear switched-capacitor networks using state charge variables. IEEE Trans. Comput. Aided Des. Integr. Circuits Syst. **4**(4), 651–661 (1985). ISSN: 1937-4151. https://doi.org/10.1109/TCAD.1985. 1270165

19. Renz, P., Deneke, N., Wicht, B.: Dynamic Modeling and Control of a Resonant Switched-Capacitor Converter with Switch Conductance Regulation, 2020 IEEE 21st Workshop on Control and Modeling for Power Electronics (COMPEL), Aalborg, Denmark, 2020, pp. 1–4, https://doi.org/10.1109/COMPEL49091.2020.9265644.

20. Leigh, J.R.: Essentials of Non-Linear Control Theory. Control, Robotics and Sensors. Peter Peregrinus on behalf of the Institution of Electrical Engineers (1983). 978-0906048962. https:// doi.org/10.1049/PBSP006E

21. Kochenburger, R.J.: A frequency response method for analyzing and synthesizing contactor servomechanisms. Trans. Am. Inst. Electr. Eng. **69**(1), 270–284 (1950). https://doi.org/10. 1109/T-AIEE.1950.5060149

22. Ackermann, J.: Robust Control – The Parameter Space Approach. Springer, London (2002). ISBN: 978-1-4471-0207-6. https://doi.org/10.1007/978-1-4471-0207-6

23. Huang, M., et al.: A fully integrated digital LDO with coarse–fine-tuning and burst-mode operation. IEEE Trans. Circuits Syst. Express Briefs **63**(7), 683–687 (2016). https://doi.org/ 10.1109/TCSII.2016.2530094

24. Huang, M., et al.: Limit cycle oscillation reduction for digital low dropout regulators. IEEE Trans. Circuits Syst. Express Briefs **63**(9), 903–907 (2016). https://doi.org/10.1109/TCSII. 2016.2534778

25. Leitner, S., et al.: Digital LDO modeling for early design space exploration. In: 2016 29th IEEE International System-on-Chip Conference (SOCC), pp. 7–12 (2016). https://doi.org/10.1109/ SOCC.2016.7905421

26. Nasir, S.B., Gangopadhyay, S., Raychowdhury, A.: All-digital low-dropout regulator with adaptive control and reduced dynamic stability for digital load circuits. IEEE Trans. Power Electron. **31**(12), 8293–8302 (2016). https://doi.org/10.1109/TPEL.2016.2519446

27. Oh, T., Hwang, I.: A 110-nm CMOS 0.7-V input transient-enhanced digital low-dropout regulator with 99.98% current efficiency at 80-mA load. IEEE Trans. Very Large Scale Integr. VLSI Syst. **23**(7), 1281–1286 (2015). https://doi.org/10.1109/TVLSI.2014.2333755

28. Kundu, S., et al.: A fully integrated digital LDO with built-in adaptive sampling and active voltage positioning using a beat-frequency quantizer. IEEE J. Solid-State Circuits **54**(1), 109–120 (2019). https://doi.org/10.1109/JSSC.2018.2870558

Chapter 7
Conclusion and Outlook

This chapter concludes the research presented in this book and gives an outlook on how resonant SC converters can be further improved by future work.

7.1 Conclusion

Li-Ion battery-powered portable devices have become part of our daily live. Long battery life is required in many applications like Internet-of-things (IoT), home automation, or wearables, which mainly operate under idle or light load conditions. Therefore, a high converter efficiency over a wide output current range is essential, in order to achieve a long battery life. DCDC converters for portable applications also require an ultra-compact and flat module size. To meet the needs of the future generations of wearable and IoT applications, the goal is to integrate all required power conversion components on one chip (SoC). Therefore, the size of the passive components must be drastically minimized.

This work focuses on fully and highly integrated hybrid DCDC converters, which are beneficial to meet the requirements for portable applications. They are a promising converter class, which merge inductive and capacitive converter approaches into one single hybrid converter. The inductor can be reduced significantly while minimizing switching losses and increasing efficiency. A study on state-of-the-art hybrid DCDC converters shows that fully integrated buck converters suffer from low efficiencies due to the high switching frequency required for inductive operation. They also often operate at low input voltages $V_{in} < 2.5\,V$ and do not support low-power operation. Conventional resonant SC converters offer high efficiencies and high power operation while significantly reducing the required switching frequency. However, the input voltage range is often limited and low-power operation is not supported.

© The Editor(s) (if applicable) and The Author(s), under exclusive license to
Springer Nature Switzerland AG 2021
P. Renz, B. Wicht, *Integrated Hybrid Resonant DCDC Converters*,
https://doi.org/10.1007/978-3-030-63944-0_7

Multi-ratio resonant operation is proposed as part of this work in order to cover the wide Li-Ion input voltage range from 3.0 V to 4.5 V. The target output voltage V_{out} is set between 1.5 V and 1.8 V and can be scaled to 1.2 V and lower. Depending on the input and output voltage, three different ratios offer a coarse control of the output voltage V_{out}. In order to cover a wide output current range from 500 µA to 120 mA, different options for the fine control are proposed depending on the size of the passives. Switch conductance regulation (SwCR) operates at the L-C resonance frequency and enables full integration of the passives. For efficient light load operation, an automatic transition into SC mode is proposed as part of this work. The converter achieves a peak efficiency of 85% with a fully integrated planar inductor and 88.5% with a 10 nH in-package inductor at power densities of 0.033 W/mm² and 0.054 W/mm². It operates at resonance switching frequencies of up to 35.5 MHz (1/2 ratio) and 47.5 MHz (2/3 and 1/3 ratios). SwCR maintains a small output voltage ripple < 30 mV. The proposed fully integrated ReSC converter is manufactured in a 130 nm BCD technology and occupies an active area of 7 mm². The second control option is called dynamic off-time modulation (DOTM) and operates with a frequency modulation of resonant current pulses. Due to the non-continuous current flow, a discrete output capacitor has to be used to achieve an acceptable output voltage ripple. In DOTM control, the converter achieves a peak efficiency of $\eta = 89\%$ with an external 10 nH inductor and 100 nF output capacitor. A peak efficiency of $\eta = 91\%$ is achieved with a higher inductance value of 39 nH.

Since the proposed resonant multi-ratio converter consists of several power switches, the implementation of the power stage is one of the major challenges for highly efficient conversion. The power switches are implemented with two stacked 1.5 V transistors due to their efficiency benefit over single 5 V transistors. Stacking of two 1.5 V transistors results in a factor of 2.3 or 3.9 lower energy loss and a factor of 1.7 or 2.8 smaller area, compared with 5 V NMOSor PMOSswitches. A new general implementation option for stacking of three low-voltage transistors for higher-voltage capability is presented, which is independent of the input voltage and can therefore be widely used in different applications. It can even be expanded to four or more stacked switches. SwCR uses segmented power switches to control the output voltage of the converter. In order to improve the loss scaling with the number of deactivated segments, this work proposes a gate decoupling circuit which leads to ~2 lower effective capacitance and significantly lower charge redistribution losses. At low output current, this results in significant efficiency improvement such as 4% at 30 mA.

For the generation of the gate overdrive voltage for each high-side switch, a bootstrapped charge pump is proposed. It generates a floating supply voltage independent from the flying reference voltage. Only one ground referred clock signal is required, while all other signals in the charge pump are self-timed. Compared to other floating charge pump or nested bootstrap solutions, no voltage drop of a diode has to be tolerated. Thus, the charge pump achieves an efficiency as high as 74%. Several floating level shifters are required for the control of the power switches. A fast capacitive level shifter transmits the clock signal to the high-side domain, which leads to an efficient high-frequency signal shifting with a propagation delay of 100 ps and an average shifting energy of 0.1 pJ. In contrast, pulsed cascode level shifters are used for the 8-bit SwCR control signals, since the controller sets the switch conductance information at a rate, which is far below the switching frequency. These level shifters are optimized to have ∼8 times lower static losses compared to the fast dynamic level shifters. This is important since the power stage consists of 58 static level shifters.

Within the scope of this research, different concepts and design aspects of integrated passives are explored. For a performance comparison, the two flying capacitors in the power stage are implemented with MIM and MOS capacitors. An additional underlying series of a highly resistively biased p-well and n-well reduces the parasitic bottom-plate capacitance of the MIM capacitor option by 35%. In SwCR control mode, the MIM capacitor option achieves up to 5% higher efficiency than the MOS capacitor option. In DOTM control mode, a 3% higher efficiency is achieved with MIM capacitors.

For the fully integrated converter option, an on-chip planar inductor is proposed. In order to determine the best trade-off between low series resistance and high inductance value, a model including major parasitic components is introduced. The model is verified by measurement results. The on-chip planar inductor has a DC resistance of $R_{DC} = 280\,\text{m}\Omega$ and a quality factor of $Q = 6$ at a switching frequency of $f = 35.5\,\text{MHz}$. In addition to the fully integrated planar inductor, two further inductor options are compared, small SMD air core inductors and microfabricated on-silicon inductors. They are both placed on top of the chip and could be co-integrated within the package in series production. With off-chip inductors, higher efficiencies can be achieved due to their better quality factors if compared to the integrated planar inductor. An efficiency improvement of up to 4% is achieved with a 10 nH SMD inductor, while a 11.9 nH microfabricated on-silicon inductor leads to an efficiency increase of up to 2.3%.

The control loops for SwCR, SC mode, and DOTM are implemented with a fast low-power mixed-signal controller. In SwCR mode, the converter uses hysteretic control, which aims to maintain the output voltage within a hysteretic window, formed by two clocked comparators. An additional third comparator with a lower threshold activates a coarse mode, which ensures a fast transient response at large load steps. At low output currents, an automatic transition in SC mode is performed, which applies frequency modulation with hysteretic discrete-time control. Fast transient response settling of <250 ns is achieved along with an output voltage drop of 5.5%. DOTM control utilizes the same controller but only operates in hysteretic discrete-time control. With an external 100 nF output capacitor, a negligible voltage droop ($\Delta V_{out} < 1\%$) is achieved. The outer control loop is responsible for setting the appropriate conversion ratio. It consists of two clocked comparators, which monitor the input voltage. A ratio controller evaluates the comparator output signals and sets the appropriate ratio. It is active in both fine control options, SwCR and SC mode as well as for DOTM.

For an optimum design of the control loop and general stability analysis, a model of the resonant SC converter and of the nonlinear mixed-signal SwCR is proposed. Due to the nonlinear behavior, stability methods such as Nyquist and

Bode cannot be applied. The resonant SC converter is modeled as a second-order system. It extends the static equivalent output resistance model by the inductor and the output capacitor. An analytic stability criterion is derived based on a harmonic balance approach together with the dual locus method. The model is verified by measurements of the proposed resonant SC converter. Unstable operating points are intentionally created to experimentally confirm that the model is well suited to predict the stability of the control loop.

In conclusion, this work meets the requirements of an ultra-compact module size, high efficiency, wide input voltage range, and wide power range in Internet-of-things and wearable applications. They are addressed by a multi-ratio hybrid resonant DCDC converter concept with on-chip inductor and capacitors, highly efficient power stage implementation, and fast and power efficient mixed-signal control.

7.2 Outlook

In this work, the air core inductor is integrated in the upper metal layers of a standard 130 nm BCD process. More benefits can be expected from future advances in inductor technologies, especially with the microfabrication of high-quality magnetic material. This could lead to higher inductance density and both lower core losses and ESR. Thus, higher efficiencies can be achieved. Together with smaller process nodes, the total solution size can be further minimized.

The resonance is defined by the actual capacitor and inductor values. In order to cope with variation of these components due to tolerances, aging, etc., frequency tuning could be implemented by observing the zero current transition of the inductor current. This information can be used for setting the required bias current for the relaxation oscillator. Techniques for zero-current detection could be adapted from dead-time control [1–3] to set the frequency during start-up of the resonant SC converter or during operation.

The benefit of resonant operation on the electromagnetic emission (EME) could be investigated in further experimental work. The sinusoidal current waveform should lead to a significantly lower EME compared to conventional hard-switching inductive converters.

SwCR control could be used as an implicit current sensor. The proposed converter already uses the counter information for a transition in low-power SC mode. During normal operation, the SwCR control loop adjusts the counter value depending on the output current. A 8-bit resolution is already achieved, but the counter value shows a strong nonlinearity, which has to be eliminated in the digital part. Afterward, the counter can be directly used as current information.

The proposed fast digital control loop for SwCR and the possible current measurement capability of the SwCR can be used in other applications like digital LDO, slope shaping, or digitally assisted analog. The introduced modeling methodology and stability analysis can also be applied in these cases.

As the presented multi-phase approach reduces the number of capacitors, it is worthwhile to further explore the benefits for resonant operation. More phases and even higher input voltages may be achievable. For more efficient coverage of a wide input voltage range, the additional conversion ratio 4/7, proposed in this work, is promising. Only one small additional flying capacitor is required. It could lead to an efficiency improvement of up to 7%.

For future power electronic applications, the high energy density of mechanical storage methods can provide an attractive alternative to the widely used L-C resonance circuit [4, 5]. Mechanical energy storage can provide at least an order of magnitude higher energy density than the electrical options [4]. They could be manufactured in a miniaturized size by MEMS technology. A mass-spring resonator is a very promising approach. Piezoelectric systems are interesting since they can drive oscillating masses and have similar expansion characteristic as springs. The Piezo effect directly provides an electromechanical conversion, required for the use in switching power converters. Very promising results can be expected in the future, but it still requires significant research on both components and circuit side.

References

1. Wittmann, J., et al.: An 18V input 10MHz buck converter with 125ps mixed-signal dead time control. IEEE J. Solid-State Circuits **51**(7), 1705–1715 (2016). ISSN: 1558-173X. https://doi.org/10.1109/JSSC.2016.2550498
2. Maderbacher, G., et al.: Automatic dead time optimization in a high frequency DCDC buck converter in 65nm CMOS. In: 2011 Proceedings of the ESSCIRC (ESSCIRC), pp. 487–490 (2011). https://doi.org/10.1109/ESSCIRC.2011.6045013
3. Abu-qahouq, J.A., et al.: Maximum efficiency point tracking (MEPT) method and digital dead time control implementation. IEEE Trans. Power Electron. **21**(5), 1273–1281 (2006). ISSN: 1941-0107. https://doi.org/10.1109/TPEL.2006.880244
4. Kyaw, P.A., Stein, A.L.F., Sullivan, C.R.: Fundamental examination of multiple potential passive component technologies for future power electronics. IEEE Trans. Power Electron. **33**(12), 10708–10722 (2018). ISSN: 1941-0107. https://doi.org/10.1109/TPEL.2017.2776609
5. Noworolski, J.M., Sanders, S.R.: An electrostatic microresonant power conversion device. In: PESC '92 Record. 23rd Annual IEEE Power Electronics Specialists Conference, vol. 2, pp. 997–1002 (1992). https://doi.org/10.1109/PESC.1992.254775

Index

© The Editor(s) (if applicable) and The Author(s), under exclusive license to   169
Springer Nature Switzerland AG 2021
P. Renz, B. Wicht, *Integrated Hybrid Resonant DCDC Converters*,
https://doi.org/10.1007/978-3-030-63944-0

Printed in the United States
by Baker & Taylor Publisher Services